AF361684

Antimicrobial Resistance and Virulence - 2nd Volume

Antimicrobial Resistance and Virulence - 2nd Volume

Editors

Manuela Oliveira
Elisabete Silva

MDPI • Basel • Beijing • Wuhan • Barcelona • Belgrade • Manchester • Tokyo • Cluj • Tianjin

Editors
Manuela Oliveira
Department of Animal Health
Faculty of Veterinary Medicine
University of Lisbon
Lisbon
Portugal

Elisabete Silva
Department of Clinics
Faculty of Veterinary Medicine
University of Lisbon
Lisbon
Portugal

Editorial Office
MDPI
St. Alban-Anlage 66
4052 Basel, Switzerland

This is a reprint of articles from the Special Issue published online in the open access journal *Antibiotics* (ISSN 2079-6382) (available at: www.mdpi.com/journal/antibiotics/special_issues/ Virulence_antibiotic).

For citation purposes, cite each article independently as indicated on the article page online and as indicated below:

LastName, A.A.; LastName, B.B.; LastName, C.C. Article Title. *Journal Name* **Year**, *Volume Number*, Page Range.

ISBN 978-3-0365-3908-9 (Hbk)
ISBN 978-3-0365-3907-2 (PDF)

Contents

About the Editors

Manuela Oliveira

Manuela Oliveira received her PhD degree from the Faculty of Veterinary University of Lisbon (FMV-ULisbon) in 2005 and is a Diplomate of the European College of Veterinary Microbiology (DipECVM). She is an Associate Professor of the Department of Animal Health of FMV-ULisbon, lecturing on Microbiology and other subjects in the Integrated Master of Veterinary Medicine and on Food Microbiology in other Master's courses. Her research interests include One Health, clinical veterinary bacteriology, bacterial biofilms, antimicrobial resistance, food safety, wildlife diseases and mycology. She has supervised and participated in several national and international research projects in these areas, authoring more than 100 peer-reviewed publications (articles, abstracts and book chapters) and 200 scientific communications in congresses (oral and poster communications). She also has experience in student supervision at several levels (PhD, MSc and LSc/BSc). Currently, she is responsible for two diagnostic laboratories of FMV-ULisbon, the Laboratory of Bacteriology and the Laboratory of Mycology.

Elisabete Silva

Elisabete Silva received her PhD degree from Instituto Superior Técnico, University of Lisbon, in 2004. She is an Auxiliary Researcher at the Department of Clinics of Faculdade de Medicina Veterinária, University of Lisbon. She also teaches practical classes on semen analysis in several veterinary species and is the quality manager of a diagnostic lab for reproductive infectious diseases, accredited by IPAQ (Portuguese Agency for Quality). Her scientific activity has been directed towards the study of bovine reproductive diseases, the identification of the molecular mechanisms that regulate sperm function, and the deciphering of Notch signaling in the regulation of reproductive/embryo development events. Her research activities also include the fields of microbiology and infectious diseases, host–pathogen interactions and antimicrobial resistance. She has participated (both as coordinator and as a team member) in several funded research projects, supervised several PhD and MSc students, and published in top international journals.

Preface to "Antimicrobial Resistance and Virulence - 2nd Volume"

The worldwide dissemination of antimicrobial-resistant bacteria, particularly those resistant to last-resource antibiotics, is a common problem to which no immediate solution is foreseen. In 2017, the World Health Organization (WHO) published a list of antimicrobial-resistant "priority pathogens", which include a group of microorganisms with high-level resistance to multiple drugs, named ESKAPE pathogens, comprising vancomycin-resistant *Enterococcus faecium* (VRE), methicillin- and vancomycin-resistant *Staphylococcus aureus* (MRSA and VRSA), extended spectrum β-lactamase (ESBL) or carbapenem-resistant *Klebsiella pneumoniae*, carbapenem-resistant *Acinetobacter baumannii*, carbapenem-resistant *Pseudomonas aeruginosa* and extended spectrum β-lactamase (ESBL) or carbapenem-resistant *Enterobacter* spp. These bacteria also have the ability to produce several virulence factors, which have a major influence on the outcomes of infectious diseases. Bacterial resistance and virulence are interrelated, since antibiotics pressure may influence bacterial virulence gene expression and, consequently, infection pathogenesis. Additionally, some virulence factors contribute to an increased resistance ability, as observed in biofilm-producing strains. The surveillance of important resistant and virulent clones and associated mobile genetic elements is essential to decision making in terms of mitigation measures to be applied for the prevention of such infections in both human and veterinary medicine, being also relevant to address the role of natural environments as important components of the dissemination cycle of these strains.

This reprint constitutes a Second Volume focusing on antimicrobial resistance and virulence, and aims to publish manuscripts that further clarify the impact of bacterial antimicrobial resistance and virulence in the three areas of the One Health triad, i.e., animal, human, and environmental health.

Manuela Oliveira and Elisabete Silva
Editors

Article

Antibiotic Susceptibility, Virulome, and Clinical Outcomes in European Infants with Bloodstream Infections Caused by Enterobacterales

Laura Folgori [1,2,*], Domenico Di Carlo [3], Francesco Comandatore [3], Aurora Piazza [4], Adam A. Witney [5], Ilia Bresesti [2], Yingfen Hsia [1,6], Kenneth Laing [5], Irene Monahan [5], Julia Bielicki [1,7], Alessandro Alvaro [3], Gian Vincenzo Zuccotti [2], Tim Planche [5], Paul T. Heath [1] and Mike Sharland [1]

1   Paediatric Infectious Disease Research Group, Institute for Infection and Immunity,
    St George's University of London, Cranmer Terrace, London SW17 0RE, UK; yhsia@sgul.ac.uk (Y.H.);
    jbielick@sgul.ac.uk (J.B.); pheath@sgul.ac.uk (P.T.H.); msharland@sgul.ac.uk (M.S.)
2   Department of Paediatrics, Vittore Buzzi Children Hospital, University of Milan, Via Lodovico Castelvetro 32,
    20154 Milan, Italy; bresesti.ilia@asst-fbf-sacco.it (I.B.); gianvincenzo.zuccotti@unimi.it (G.V.Z.)
3   Paediatric Clinical Research Centre "Romeo and Enrica Invernizzi", Department of Biosciences,
    University of Milan, Via Giovanni Battista Grassi 74, 20157 Milan, Italy; domenico.dicarlo@unimi.it (D.D.C.);
    francesco.comandatore@unimi.it (F.C.); alessandro.alvaro@unimi.it (A.A.)
4   Clinical-Surgical, Diagnostic and Pediatric Sciences Department, Unit of Microbiology and Clinical
    Microbiology, University of Pavia, Corso Str. Nuova 65, 27100 Pavia, Italy; aurora.piazza@unipv.it
5   Institute of Infection and Immunity, St George's University of London, Cranmer Terrace,
    London SW17 0RE, UK; awitney@sgul.ac.uk (A.A.W.); klaing@sgul.ac.uk (K.L.); imonahan@sgul.ac.uk (I.M.);
    timothy.planche@stgeorges.nhs.uk (T.P.)
6   School of Pharmacy, Queen's University, 97 Lisburn Rd., Belfast BT9 7BL, UK
7   Pediatric Pharmacology and Pharmacometrics, University Children's Hospital (UKBB),
    University Hospital Basel, Spitalstrasse 33, 4056 Basel, Switzerland
*   Correspondence: folgori.laura@asst-fbf-sacco.it; Tel.: +44-20-87254851

Citation: Folgori, L.; Di Carlo, D.; Comandatore, F.; Piazza, A.; Witney, A.A.; Bresesti, I.; Hsia, Y.; Laing, K.; Monahan, I.; Bielicki, J.; et al. Antibiotic Susceptibility, Virulome, and Clinical Outcomes in European Infants with Bloodstream Infections Caused by Enterobacterales. *Antibiotics* **2021**, *10*, 706. https://doi.org/10.3390/antibiotics10060706

Academic Editors: Manuela Oliveira and Elisabete Silva

Received: 12 May 2021
Accepted: 8 June 2021
Published: 11 June 2021

Publisher's Note: MDPI stays neutral with regard to jurisdictional claims in published maps and institutional affiliations.

Abstract: Mortality in neonates with Gram-negative bloodstream infections has remained unacceptably high. Very few data are available on the impact of resistance profiles, virulence factors, appropriateness of empirical treatment and clinical characteristics on patients' mortality. A survival analysis to investigate 28-day mortality probability and predictors was performed including (I) infants <90 days (II) with an available Enterobacterales blood isolate with (III) clinical, treatment and 28-day outcome data. Eighty-seven patients were included. Overall, 299 virulence genes were identified among all the pathogens. *Escherichia coli* had significantly more virulence genes identified compared with other species. A strong positive correlation between the number of resistance and virulence genes carried by each isolate was found. The cumulative probability of death obtained by the Kaplan-Meier survival analysis was 19.5%. In the descriptive analysis, early age at onset, gestational age at onset, culture positive for *E. coli* and number of classes of virulence genes carried by each isolate were significantly associated with mortality. By Cox multivariate regression, none of the investigated variables was significant. This pilot study has demonstrated the feasibility of investigating the association between neonatal sepsis mortality and the causative Enterobacterales isolates virulome. This relationship needs further exploration in larger studies, ideally including host immunopathological response, in order to develop a tailor-made therapeutic strategy.

Keywords: infant; newborn; bacteremia; Gram-negative bacteria; drug resistance; microbial; virulence factors; mortality

1. Introduction

Mortality in neonates with Gram-negative bloodstream infections (GN-BSIs) has remained unacceptably high. Appropriate empirical treatment is considered crucially important in reducing mortality. However, despite the improvement in neonatal care, the fatality rate in babies with GN-BSIs remains around 15–20%, also during the emergence of antimicrobial resistance (AMR) [1–6].

Very few data are available on the impact of different treatment regimens on clinical outcome in neonates with GN-BSIs. Previous studies conducted in both adults and children showed conflicting results on the impact of resistance profiles, appropriateness of empirical treatment and clinical characteristics on patients' mortality [7–15]. In the last decade, the implementation of modern bioinformatics to assist next-generation sequencing data analysis greatly improved the knowledge on the genetic characterization of pathogenic strains that may serve as target for new therapies [16]. There has been a growing body of evidence about the role of virulence factors (VFs) in the pathogenesis of invasive infections. Enterobacterales employ many strategies to enhance invasiveness, overcome host defenses and cause infections. Different strains can use alternative VFs with similar functions during the infection process, with this plasticity leading to unique combinations of such factors [17]. Some VFs are disease-specific whereas others seem to play different roles in different types of infection [18]. This plasticity enables pathogenic strains to colonize and infect different tissues and hosts. Some major classes of VFs, such as capsule, siderophores and fimbriae, have been characterized well [18]. However, several other factors were recently identified and have yet to be defined to fully understand their mechanisms of action and clinical significance (summarized in Table 1). To achieve this goal, a genomic approach can be used to identify genes encoding specific virulence determinants. In adults, several genes present in the great majority of bacteremic strains and involved in virulence have been identified [19–21]. These are presumably essential for the infection process. However, the specific VFs that are relevant in causing neonatal GN-BSIs are not well defined yet, partly due the variability among the few studies available so far [2]. These have mostly been conducted on neonates and children with *Escherichia coli* bacteremia, with virtually no data available on other Enterobacterales [22–25]. Also, in the recent years, several data have been published investigating a potential role of the bacterial virulome, defined as the set of genes contributing to the bacterial virulence, in determining the outcome of patients with both Gram-positive and GN-BSIs. Again, these studies demonstrate a mixed picture reporting a significant correlation between virulence factors and mortality in both children and adults [2,19,26–29].

Table 1. Main virulence factors in Enterobacterales.

Category	Sub-Category	Function	Genes	Pathogens
Adherence	Anti-aggregation protein-dispersin	Bound to the outer membrane, assisting dispersion across the surface by overcoming electrostatic attraction between fimbriae and bacterial surface	*aap*	*Escherichia coli*
	Adhesins	Cell-surface components that allow bacteria to attach to host cells or to surfaces	*afa, dra, fde*	*Escherichia coli*
	Fimbriae	Major adhesive structures in biofilm formation and binding to abiotic surfaces	*agg, bcf, csg, daa, fim, foc, lpf, mrk, paa, pap, sfa, yag.ecp, ykg.ecp*	*Escherichia coli* *Enterobacter* spp. *Klebsiella* spp.
	Intimin	Outer membrane protein needed for intimate adherence	*eae, tir*	*Escherichia coli*
	Zinc metalloprotease	Contributes to intimate adherence to host cells	*stc*	*Escherichia coli*

Table 1. *Cont.*

Category	Sub-Category	Function	Genes	Pathogens
Bacterial metabolism	Allantoin Metabolism	Enzymes involved in degradation of allantoin	*all* A-D	*Klebsiella* spp. *Escherichia coli* *Enterobacter* spp.
	Transcription factors	DNA-binding transcriptional activator/repressor	*all* R-S	*Klebsiella* spp. *Escherichia coli* *Enterobacter* spp.
	Bacterial survival promoters	Methionine aminopeptidase	*map*	*Escherichia coli*
		Magnesium transporter	*mgt*	*Klebsiella* spp. *Escherichia coli* *Enterobacter* spp.
		Toll-like receptor and MyD88-specific signalling inhibitor	*tcp*	*Escherichia coli*
Capsule	Capsule	Extracellular polysaccharide matrix that envelops the bacteria, prevents phagocytosis, hinders the bactericidal action of antimicrobial peptides, blocks complement components	*cps, gal, glf, gnd, gtr, kfo, kps, man, rcs, rmp, ugd, wca, wza, wzi, wzm, wzt*	*Klebsiella* spp. *Escherichia coli* *Enterobacter* spp.
	Lipopolysaccharide	Component of the outer leaflet of the cell membrane of all Gram-negative bacteria (GNB) which protects against humoral defences	*lpx, waa, wbb*	All GNB
Cell invasion	Arylsulfatase	Penetration of the blood-brain barrier	*asl*	*Escherichia coli*
	Outer membrane porin A	Adherence to epithelial cells, translocation into epithelial cells nucleus, induction of epithelial cell death, biofilm formation, binding to factor H to allow bacteria to develop serum-resistance	*omp*A	*Escherichia coli* *Enterobacter* spp. *Klebsiella* spp.
	Invasion protein A	Cell invasion into the host tissues	*ibe*A	*Escherichia coli*
Iron metabolism	Siderophores-Hemin uptake	Enable using of Fe from haemoglobin in the host system	*chu*	*Escherichia coli* *Enterobacter* spp.
	Siderophores-Enterobactin	Mediation of iron acquisition, obstacole macrophages antimicrobial responses	*ent, fep, fes*	All GNB
	Siderophores-Yersiniabactin	Can solubilise iron bound to host binding proteins and transport it back to the bacteria	*fyu, irp, ybt*	*Escherichia coli* *Klebsiella* spp.
	Siderophores-Salmochelin	Siderophore receptor, use of Fe ions obtained from the body host	*iro*	*Escherichia coli* *Enterobacter* spp. *Klebsiella* spp.
	Siderophores-Aerobactin	Acquisition of $Fe^{2+/3+}$ in the host system	*iuc, iut*	*Escherichia coli* *Klebsiella pneumoniae*
	Heme/haemoglobin transport protein and receptor	Cell survival	*shu*	*Escherichia coli* *Enterobacter* spp.
Motility and chemotaxis	Chemotaxis	Bacterial movement in response to a chemical stimulus	*che, mot*	*Escherichia coli* *Enterobacter* spp.
	Flagella	Motility organelle, function as adhesins	*flg, flh, fli*	*Escherichia coli* *Enterobacter* spp.
Pumps	Pumps	Efflux pump implicated in both virulence and resistance to antibiotics	*acr*	*Escherichia coli* *Enterobacter* spp. *Klebsiella* spp.

Table 1. *Cont.*

Category	Sub-Category	Function	Genes	Pathogens
Secretion system factor	Type I secretion system protein (T1SS)	Enables pathogens to inject effector proteins into host cells	*hly*	*Escherichia coli*
	Type II secretion system protein (T2SS)	Enables pathogens to inject effector proteins into host cells	*exe, gsp*	*Escherichia coli* *Klebsiella* spp.
	Type III secretion system (T3SS)	Enables pathogens to inject effector proteins into host cells	*ces, esc*	*Escherichia coli*
	Type VI secretion system (T6SS)	Enables pathogens to inject effector proteins into host cells	*clp*V.*tss*H, *dot*U.*tss*L, *hcp, hsi*B1.*vip, icm*F.*tss*	*Escherichia coli* *Enterobacter* spp. *Klebsiella* spp.
Toxins	Colibactin	Genotoxin causing genomic instability in eukaryotic cells by induction of double-strand breaks in DNA	*clb*	*Escherichia coli* *Klebsiella pneumoniae*
	T3SS effector	Cytoskeletal rearrangements	*esp*	*Escherichia coli*
	Hemolysin A	Creating of pores in membranes of host cells (cell lysis)	*hly*	*Escherichia coli*

Clarifying the role of the main determinants leading to adverse outcomes could help to define targeted interventions to decrease mortality. With this study, we aimed to investigate potential associations between patient characteristics, pathogen characteristics and antibiotic treatment regimen on the clinical outcome of neonates/infants affected by culture-proven GN-BSIs.

2. Results

2.1. Demographic and Clinical Data

Overall, 87 infants from six European countries (the United Kingdom: 49, Estonia: 21, Greece: 7, Italy: 7, Lithuania: 2, Spain: 1) between 2010–2015 fulfilled our inclusion criteria and were included in the study. Forty patients were retrieved from the neonatal infection surveillance network (NeonIN) study [30], 38 from NeoMero [31], and 9 from the Collaborations for Leadership in Applied Health Research and Care (CLAHRC) study [32]. Patients, pathogens and treatment characteristics of the included episodes are summarised in Table 2. At the BSI onset, the median age of the selected neonates was 15.2 days (interquartile range (IQR) 6.7–31), with a median gestational age (GA) of 33 weeks (IQR 28–37). Forty-nine out of 87 babies (56%) had a central line in situ at the episode onset. A total of 37 different antibiotic regimens have been reported among the 87 patients in the first 48 h of treatment.

Table 2. Demographic and clinical characteristics of included patients.

Variable	Overall, n = 87 (%)
Gender	
Male	43 (49)
Female	44 (51)
Age at the onset (days, median (IQR))	15.2 (6.7–31)
Gestational age category (weeks of GA)	
<28 0/7	37 (42)
28 0/7–31 6/7	21 (24)
32 0/7–33 6/7	8 (9)
34 0/7–36 6/7	8 (9)

Table 2. *Cont.*

Variable	Overall, $n = 87$ (%)
37 0/7–38 6/7	7 (8)
39 0/7–40 6/7	6 (7)
Birth weight category (grams)	
>=2500	20 (23)
1500–<2500	15 (17)
1000–<1500	15 (17)
<1000	37 (42)
Small for Gestational Age (SGA)	
Yes	11 (13)
No	76 (87)
Underlying conditions	
Yes	51 (59)
No	36 (41)
Gestational age (weeks) at onset, median (IQR)	33 (28–37)
Isolated organism	
Escherichia coli	36 (41)
Enterobacter cloacae	18 (21)
Klebsiella pneumoniae	11 (13)
Klebsiella oxytoca	7 (8)
Serratia marcescens	7 (8)
Enterobacter asburiae	3 (3)
Enterobacter aerogenes	2 (2)
Serratia liquefaciens	1 (1)
Enterobacter kobei	1 (1)
Proteus mirabilis	1 (1)
First 48-h antibiotic treatment *	
Aminoglycosides anti-bacterials	25 (29)
Beta-lactam anti-bacterials, penicillins	26 (30)
Other anti-bacterials	9 (10)
Other beta-lactam anti-bacterials	23 (26)
Quinolone anti-bacterials	4 (5)
First 48-h treatment concordance with the anti-biogram	
Concordant	81 (93)
Discordant	6 (7)
Multidrug resistant **	
No	61 (70)
Yes	26 (30)
Number of classes of resistance genes per isolate, median (IQR)	5 (4–5)
Number of classes of virulence genes per isolate, median (IQR)	7 (7–9)

* Coded according to the WHO ATC/DDD (Anatomical Therapeutic Chemical/Defined Daily Dose) Index 2020 at the 4th level. ** According to Magiorakos, A.P., *Clin. Microbiol. Infect.* **2012** Mar, *18*(3), 268–281.

2.2. Microbiological Data

The species IDs identified on matrix-assisted laser desorption/ionization time-of-flight (MALDI-TOF) mass spectrometry were all confirmed by sequencing. The isolate-specific accession numbers are indicated in Supplementary Table S1. The most frequently isolated pathogen was *E. coli* followed by *Enterobacter* spp. and *Klebsiella* spp. (Table 2). The percentage of multidrug-resistance (MDR) isolates was 30% (26/87). Based on interpretation of the in vitro susceptibility profile, 16/87 (18%) were suspected of producing an extended-spectrum beta lactamase (ESBL) enzyme, and only one isolate (*Klebsiella pneumoniae*) was resistant to carbapenems. Susceptibilities of single species to the investigated antibiotics are presented in Table 3.

Table 3. Percentages of susceptibility for the isolated pathogens.

Pathogen (*n*)	AMK *	AMP	AMC	CRO	CAZ	ATM	CIP	GEN	MEM	TZP	SXT
Escherichia coli (36)	97	42	67	94	97	94	89	86	100	92	64
Enterobacter cloacae (18)	94	0	0	56	56	83	100	67	100	94	89
Klebsiella pneumoniae (11)	55	0	46	46	55	64	91	64	91	64	64
Klebsiella oxytoca (7)	100	0	71	86	86	100	100	71	100	100	86
Serratia marcescens (7)	100	0	0	86	100	86	86	100	100	86	100
Enterobacter asburiae (3)	100	0	0	100	100	100	100	33	100	100	100
Enterobacter aerogenes (2)	50	0	0	50	50	50	100	50	100	50	100
Serratia liquefaciens (1)	100	100	100	100	100	100	100	100	100	100	100
Enterobacter kobei (1)	100	100	0	100	100	100	100	100	100	100	100
Proteus mirabilis (1)	100	0	100	100	100	100	100	0	100	100	0

AMK: amikacin, AMP: ampicillin, AMC: amoxicillin-clavulanate, CRO: ceftriaxone, CAZ: ceftazidime, ATM: aztreonam, CIP: ciprofloxacin, GEN: gentamicin, MEM: meropenem, TZP: piperacillin-tazobactam, SXT: trimethoprim-sulphametoxazole * Proportion of isolates resistant to the antibiotic.

A total of 50 different sequence types (STs) were found, with ST131 and ST90 as the most frequent in *E. coli* and *Enterobacter cloacae*, respectively. The median number of classes of resistance genes carried per isolate was 5 (IQR 4–5) whereas the median number of classes of virulence genes was 7 (IQR 7–9). Twenty-five isolates harbored blaTEM-type genes, two non-*Klebsiella* spp. strains the blaSHV-type determinant, and two *E. coli* strains the blaCTX-M-type genes. One *K. pneumoniae* ST17 carried blaVIM-12 gene, and two *Enterobacter asburiae* (ST484) the mcr-9 determinant.

Overall, the genome sequencing identified 299 different virulence genes among all the pathogens. There was a strong positive correlation between the number of resistance and virulence genes carried by each isolate (*Rho* = 0.79; *p* = 0.001) (Figure 1).

E. coli strains showed the highest mean number of virulence genes (105 vs. <65 in the other species overall), mainly those involved in fimbriae production (*p* < 0.0001) (Table 4). The genes that were more frequently carried by the isolates are summarized in Table 5. Among the most represented strains (*E. coli*, *E. cloacae*, *K. pneumoniae*, *K. oxytoca*) the following virulence genes were carried by all the isolates: *gal*, *gnd*, *rcs* (capsule); *ompA* (cell invasion); *ent*, *fep* (iron metabolism); *acr* (pumps). Some genes were shown to be strain-specific. Among them, the adhesion *mrk* gene cluster was sequenced in all *Klebsiella* spp. isolates as well as the secretion system's *exe* and *impA.tss* genes. Conversely, genes encoding for motility and chemotaxis proteins (*che*, *flg* and *fli*) were only carried by *E. coli* and *Enterobacter* spp. The *clb*, *esp* and *hly* genes encoding for toxin proteins were sequenced only in *E. coli* strains. One hypervirulent *K. pneumoniae* with a hypermucoviscous phenotype harboring the *rmpA* and *wca* genes was found [33].

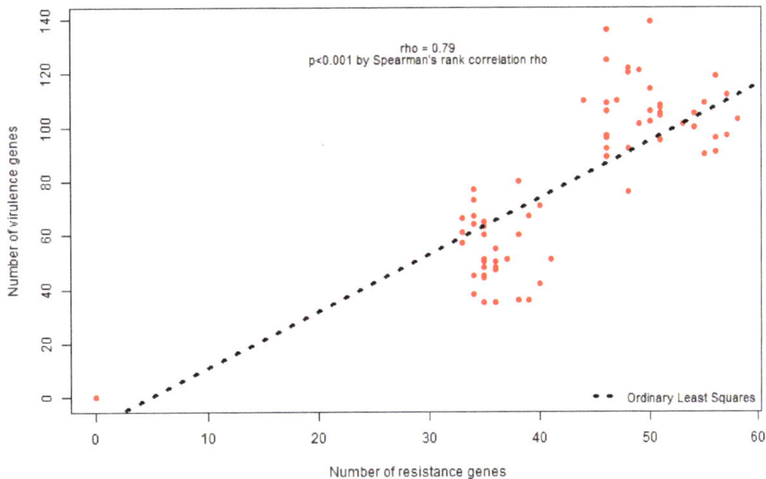

Figure 1. Spearman's rank correlation between number of resistance and virulence genes.

Table 4. Median number of classes of virulence genes carried per bacterial strain.

Pathogen (*n*)	Virulence Factors Category (Median Number of Genes Carried)								
	Adherence	Bacterial Metabolism	Capsule	Cell Invasion	Iron Metabolism	Motility and Chemotaxis	Pumps	Secretion System Factor	Toxins
Escherichia coli (36)	31	2	5.5	2	37	8	2	13	6.5
Enterobacter cloacae (18)	5	0	7	1	16	7	2	11	0
Klebsiella pneumoniae (11)	24	0	14	1	12	0	2	13	0
Klebsiella oxytoca (7)	15	2	9	1	22	0	2	10	0

Table 5. Number and percentage of the most represented virulence genes.

Virulence Gene	*Escherichia coli* (36)	*Enterobacter cloacae* (18)	*Klebsiella pneumoniae* (11)	*Klebsiella oxytoca* (7)
	N (%)	*N* (%)	*N* (%)	*N* (%)
Adherence				
csg	36 (100)	18 (100)	0	0
fde	36 (100)	0	0	0
fim	36 (100)	0	10 (91)	7 (100)
mrk	0	2 (11)	11 (100)	7 (100)
pap	26 (72)	0	0	0
yag.ecp	36 (100)	0	11 (100)	7 (100)
ykgK.ecp	35 (97)	0	11 (100)	0

Table 5. *Cont.*

Virulence Gene	*Escherichia coli* (36)	*Enterobacter cloacae* (18)	*Klebsiella pneumoniae* (11)	*Klebsiella oxytoca* (7)
	N (%)	*N* (%)	*N* (%)	*N* (%)
Bacterial metabolism				
all	36 (100)	0	0	3 (43)
mgt	0	0	0	7 (100)
Capsule				
cps	0	0	10 (91)	0
gal	36 (100)	18 (100)	11 (100)	7 (100)
gif	0	0	6 (54)	7 (100)
gnd	36 (100)	18 (100)	11 (100)	7 (100)
kps	33 (92)	0	0	0
man	0	18 (100)	7 (64)	2 (29)
rcs	36 (100)	18 (100)	11 (100)	7 (100)
ugd	2 (6)	18 (100)	11 (100)	7 (100)
Cell invasion				
asl	36 (100)	0	0	1 (14)
ompA	36 (100)	18 (100)	11 (100)	7 (100)
Iron metabolism				
chu	35 (97)	6 (33)	0	0
ent	36 (100)	18 (100)	11 (100)	7 (100)
fep	36 (100)	18 (100)	11 (100)	7 (100)
fes	36 (100)	0	11 (100)	7 (100)
fyu	35 (97)	0	2 (18)	5 (71)
iro	11 (31)	11 (61)	11 (100)	0
irp	35 (97)	0	2 (18)	5 (71)
iuc	25 (69)	0	1 (9)	0
iut	25 (69)	0	1 (9)	0
Motility and chemotaxis				
che	36 (100)	18 (100)	0	0
flg	36 (100)	18 (100)	0	0
fli	36 (100)	18 (100)	0	0
Pumps				
acr	36 (100)	18 (100)	11 (100)	7 (100)
Secretion system factor				
clpV.tssH	0	13 (72)	11 (100)	3 (43)
dotU.tssL	0	13 (72)	11 (100)	4 (57)
exe	0	0	11 (100)	7 (100)
gsp	35 (97)	0	0	0

Table 5. *Cont.*

Virulence Gene	*Escherichia coli* (36)	*Enterobacter cloacae* (18)	*Klebsiella pneumoniae* (11)	*Klebsiella oxytoca* (7)
	N (%)	*N* (%)	*N* (%)	*N* (%)
hcp	21 (58)	14 (78)	11 (100)	7 (100)
hsiB1.vip	0	18 (100)	0	0
icmF.tss	0	11 (61)	10 (91)	4 (57)
impA.tss	0	0	11 (100)	0
vasE.tssK	0	12 (67)	11 (100)	4 (57)
vip.tss	21 (58)	15 (83)	11 (100)	7 (100)
ybd	0	18 (100)	10 (91)	7 (100)
Toxins				
clb	7 (19)	0	1 (9)	0
esp	15 (42)	0	0	0

2.3. Determinants of 28-Day Case-Fatality

The cumulative probability of death obtained by the Kaplan-Meier survival analysis was 19.5% with the greater percentage of deaths happening in the first week. In the descriptive analysis, early age at onset ($p = 0.002$), culture positive for *E. coli* ($p = 0.029$), number of classes of virulence genes carried per isolate ($p = 0.022$) and GA (weeks) at the onset ($p = 0.003$) were significantly associated with mortality (Table 6). By Cox multivariate regression, none of the investigated variables was significant (Table 7).

Table 6. Descriptive analysis of potential association between variables and mortality.

Variable	Alive, *n* = 69 (%)	Died, *n* = 18 (%)	*p*-Value
Male	31 (45)	12 (67)	0.168
Female	38 (55)	6 (33)	
Age at the onset (days, median (IQR))	19.1 (8.8–35)	7.1 (3.3–9.8)	**0.002**
Gestational age category (weeks of GA)			
<28 0/7	27 (39)	10 (56)	0.323
28 0/7–31 6/7	17 (25)	4 (22)	1.000
32 0/7–33 6/7	7 (10)	1 (6)	1.000
34 0/7–36 6/7	8 (12)	0 (0)	0.197
37 0/7–38 6/7	5 (7)	2 (11)	0.631
39 0/7–40 6/7	5 (7)	1 (6)	1.000
Birth weight category (grams)			
>=2500	17 (25)	3 (17)	0.754
1500–<2500	13 (19)	2 (11)	0.727
1000–<1500	13 (19)	2 (11)	0.727
<1000	26 (38)	11 (61)	0.128
Small for Gestational Age (SGA)			
Yes	8 (12)	3 (17)	0.690
No	61 (88)	15 (83)	

Table 6. *Cont.*

Variable	Alive, n = 69 (%)	Died, n = 18 (%)	*p*-Value
Underlying conditions			
Yes	39 (56)	12 (67)	0.610
No	30 (43)	6 (33)	
Gestational age (weeks) at onset, median (IQR)	33 (30–38)	27 (25–33)	**0.003**
Isolated organism			
Escherichia coli	24 (35)	12 (67)	**0.029**
Enterobacter spp.	20 (29)	4 (22)	0.769
Klebsiella spp.	16 (23)	2 (11)	0.342
Serratia spp / *Proteus mirabilis*	9 (13)	0 (0)	0.194
First 48-h antibiotic treatment *			
Aminoglycosides antibacterials	19 (27)	6 (33)	0.848
Beta-lactam antibacterials, penicillins	22 (32)	4 (22)	0.611
Other antibacterials	7 (10)	2 (11)	1.000
Other beta-lactam antibacterials	19 (27)	4 (22)	0.770
Quinolone antibacterials	2 (3)	2 (11)	0.188
First 48-h treatment concordance with the antibiogram			
Concordant	64 (93)	17 (94)	1.000
Discordant	5 (7)	1 (6)	1.000
Multidrug resistant			
No	47 (68)	14 (78)	1.000
Yes	22 (32)	4 (22)	1.000
Number of classes of resistance genes per isolate, median (IQR)	5 (4–5)	5 (4–5)	0.203
Number of classes of virulence genes per isolate, median (IQR)	7 (7–9)	9 (8–9)	**0.022**

* coded according to the WHO ATC/DDD Index 2020 at the 4th level.

Table 7. Multivariate regression analysis on the 28-day mortality predictors.

Variable	N	HR * (L.95–U.95)	*p*-Value
Age at the onset (days)	87	0.97 (0.93–1.01)	0.125
N. classes of virulence genes	87	1.35 (0.92–1.97)	0.128
Gestational Age (weeks) at onset	87	0.91 (0.83–1)	0.056

* Hazard Ratio.

3. Discussion

This study included 87 European neonates and infants younger than 90 days with GN-BSIs due to Enterobacterales. Overall, 299 virulence genes were identified in these pathogens. Among the different organisms, *E. coli* had significantly more virulence genes identified compared with other species. *Gal, gnd, rcs, ompA, ent, fep,* and *acr* virulence genes were identified from all the pathogens, likely being essential for the infection process. Conversely, some other genes were shown to be strain-specific. A strong positive correlation between the number of resistance and virulence genes carried by each isolate was found. By survival analysis, the 28-day probability of death was 19.5%. In the descriptive analysis, early age at onset, GA at the onset, culture positive for *E. coli* and number of classes of virulence genes carried by each isolate were significantly associated with mortality whereas discordant therapy was not related to mortality. By Cox multivariate regression, none of the investigated variables was significant.

Many studies have been conducted in both adults and neonates trying to define the main determinants of mortality in patients with GN-BSIs. AMR has been broadly investigated in the adult population, with the majority of studies reporting a significant association between multiple resistance to antibiotics and patients mortality [7,34–36]. Some large cohorts have not found a clear correlation between AMR and adverse outcome [8,12,13]. Data from this small study did not confirm a significant impact of resistance profile on neonatal mortality [5].

In recent years, an increasing number of studies are being conducted trying to investigate the impact of virulence genes on the outcome of patients with GN-BSIs. Among them, *E. coli* was the most frequently investigated pathogen followed by *K. pneumoniae*. Independent risk factors associated with 30-day mortality among adult patients with ESBL-producing *E. coli* bacteremia included siderophores *iroN* and *iss* positivity [21], the siderophore *fyu*A gene, and the presence of the afimbrial adhesin *afa* gene [19,37]. In a large prospective study investigating the main determinants for adverse outcome in patients with *K. pneumoniae* BSIs, the cytotoxicity *pks* gene cluster carriage by causative strains was an independent risk factor for 30-day mortality when accompanied by MDR [38]. Lastly, the siderophore-related *iut*A gene was found to be an independent predictor of the 30-day mortality in *K. pneumoniae* bacteremia [39]. However, almost all of these studies were conducted with pre-selected virulence genes searched by polymerase chain reaction (PCR) rather than sequencing the entire bacterial virulome. This led to a wide heterogeneity, with each group analyzing different genes and hampering the comparison of the results. Very little data have been published on the relationship between bacterial virulence factors and BSI mortality in children, and almost all in patients with *E. coli* bacteremia. In a prospective cohort of 43 septic neonates, the adhesin *hek/hra* gene was found to be significantly more frequent in isolates from newborns who died than in isolates from survivors [2]. On the other hand, in a cohort of children 0–17 years old (median age 2.4 months), none of the 20 virulence factors tested by PCR was found to correlate with sepsis severity [26]. The Burden of Antibiotic Resistance in Neonates from Developing Societies (BARNARDS) study was conducted to assess the burden of AMR in neonates in seven low-middle income countries [6]. In this study, Gram-negative (GN) pathogens from neonatal sepsis were isolated and characterized through whole genome sequencing (WGS) for resistance and virulence genes. The number of virulence genes carried by each isolate through a virulence score was used. The results obtained suggested that yersiniabactin and/or aerobactin/salmochelin virulence genes may be involved in a more rapid onset and mortality. However, the inability to follow up all neonates and additional local factors likely to contribute to patient's death hampered the authors' capacity to attribute mortality singularly to the presence/absence of genomic traits.

Our results showed a strong positive correlation between the number of resistance and virulence genes carried by each isolate. Many studies have been conducted on either AMR or virulence. However, the biological effect and connection between these two factors are of particular importance [40,41]. Indeed, a negative or positive relationship can be found among them. Enhanced virulence or AMR frequently has been reported to have a fitness cost on bacteria but their relationship changes according to different bacterial species, the resistance and virulence genes involved and the host's immune system [42,43]. Some antibiotics, such as ceftazidime, cefotaxime and quinolones, have been reported to enhance the increase of the deletion and transposition of DNA regions that are specific for VFs [42]. By contrast, a positive correlation has been shown between AMR and virulence with the use of other antibiotics. In particular, uropathogenic strains of *E. coli* carrying the *bla*CTX-M-15 resistance gene also harbored more *col*V, *col*E2-E9, *col*Ia-Ib, *hly*A, and *csg*A genes as well as the *bla*OXA-2 beta lactamase was correlated with increased expression of *col*M, *col*B, *col*E, and *crl* genes [44]. Prophages are another mechanism that has been shown to be involved in both virulence and resistance diffusion through the spread of toxins to other resistant strains [45]. Porins and biofilm also play an important role in the relationship between virulence and resistance, with the first acting as a channel controlling the entrance of both antibiotics and VFs into the pathogens (i.e., the *OmpC* gene) and the second favoring antimicrobial treatment tolerance and infection persistence at the same time also increasing the transfer of resistance and virulence genes among the cells [46]. Lastly, an enhancement in both resistance and virulence characteristics of the pathogens can occur through mobile genetic elements like plasmids. These self-replicating extra-chromosomal elements are capable of transferring among different bacteria while carrying virulence and resistance genes [47]. This mechanism is independent of any antibiotic pressure.

There are several limitations in the study design, due to both confounding factors and heterogeneity of our sample. Firstly, the study is not powered to demonstrate a significant correlation between the presence of single resistance/virulence genes and neonates' clinical outcome; this required us to analyze resistome and virulome by grouping genes according to the class. Moreover, we are well aware of the rapid dynamics of bacterial genetics, and the selection of a single time point isolate can cause bias and affect the results. At the same time, considering the possible implication of heteroresistance, the selection of certain colonies in the first place could have potentially missed relevant isolates. We found a significantly higher number of virulence genes in *E. coli* isolates compared with other species; this could be due to an over representation of *E. coli* genes in the virulence database (VFDB). Different pathogens can have different impacts on neonates with sepsis, and pooling data on multiple strains could alter the results. Lastly, neonates and infants represent a broadly heterogeneous population, being characterized by different gestational ages, birth weight, underlying conditions, and risk factors. Despite these limitations, this was the first study to correlate virulence factors of non-*E. coli* Enterobacterales with mortality in septic neonates. Although a number of studies have tried to correlate single virulence genes with Gram-negative sepsis outcome, virtually none of them have investigated the role of the entire virulome in causing mortality. Considering the huge number of virulence genes involved and the limited sample size, we presented a potential way to use the entire body of information gained from the WGS by grouping genes as the number of classes of virulence genes involved. Having access to a very complete dataset including patient-level treatment, outcome data, and isolates available for a comprehensive genetic characterization, we believe that our data provide new and relevant information on the molecular picture of GN pathogens causing neonatal BSIs.

4. Materials and Methods

4.1. Study Design and Data Source

A case series of neonates with clinical sepsis and microbiologically confirmed GN-BSIs was constructed from three separate studies: the NeonIN [30], the NeoMero clinical trial [31], and the CLAHRC [32]. The NeonIN is a multinational network which prospectively collect data on neonatal infections from neonatal units. The detailed data procedures have been previously described [48]. NeoMero was a European-based randomized controlled trial to compare the efficacy of meropenem with standard of care (ampicillin + gentamicin or cefotaxime + gentamicin) in the treatment of late onset neonatal sepsis. CLAHRC is a United Kingdom-based prospective cohort study to collect data for patients with GN-BSIs in three hospitals in South London which aims to characterize clinical management of patients with GN-BSIs in all ages and identify potential risk factors associated with 28-day treatment outcomes.

4.2. Selection Criteria, Available Data and Definitions

Patients with a microbiologically-confirmed diagnosis of GN-BSIs due to Enterobacterales were selected from the above studies among European Neonatal Intensive Care Units between 2010–2015. Inclusion criteria were (I) age between 0–90 days, (II) Enterobacterales blood isolate available for the sequencing, (III) patient-level clinical and treatment data, and (IV) 28-day clinical outcome data.

Data available for the analysis included demographic, risk factors, clinical characteristics, antibiotic treatment, susceptibility results, resistome and virulome of the isolated bacteria, and 28-day treatment outcome.

The date of the blood culture was considered as the day of the sepsis onset. The MDR of the isolates was defined according to Magiorakos A.-P. et al. as acquired non-susceptibility to at least one agent in three or more antimicrobial categories [49]. The first 48-h antibiotic treatment was defined as concordant or discordant based on the inclusion of at least an active drug against the blood isolate. The evaluated outcome was defined as 28-day case-fatality (patient alive/dead).

4.3. Microbiological Methods

Bacterial isolates from blood were cultured from frozen stocks (-80 °C) on blood agar plates. Species were identified by MALDI-TOF mass spectrometry (Bruker, Karlsruhe, Germany). Antibiotic susceptibility profiles were obtained according to the European Committee on Antimicrobial Susceptibility Testing (EUCAST) 2019 Clinical Breakpoints with disk diffusion tests for the following antibiotics: amikacin, ampicillin, amoxicillin-clavulanate, ceftriaxone, ceftazidime, aztreonam, ciprofloxacin, gentamicin, meropenem, piperacillin-tazobactam, and trimethoprim-sulphametoxazole. To facilitate the analyses, isolates that were defined as having increased exposure were classified as non-susceptible. The isolates included in the study were subjected to WGS using the Illumina MiSeq platform (Illumina, San Diego, CA, USA), with paired-end runs of 2×300 bp, after Nextera XT library preparation. The obtained reads were assembled using SPAdes [50]. For each genome, we determined the ST using an in-house script (available upon request) and the Multilocus sequence typing (MLST) schemes and gene alleles sequences available on PubMLST (pubmlst.org). The isolates were further characterized at the genomic level with the identification of resistant and virulence genes using ABRicate (Seemann T, Abricate, Github https://github.com/tseemann/abricate, accessed on: 11 November 2019) and the following databases: The Comprehensive Antibiotic Resistance Database (CARD) [51] and Resfinder [52] for the resistance genes and VFDB [53] for the virulence genes.

4.4. Statistical Analysis

Qualitative variables were summarized by absolute frequencies and percentages, and quantitative variables by median and IQR. A descriptive analysis was conducted with the potential association between variables and outcome of interest evaluated by chi-squared or Fisher's exact test as more appropriate for qualitative variables, and Mann–Whitney or t-test as appropriate for quantitative variables. A Spearman's rank correlation was calculated to evaluate the correlation between the number of resistance genes and virulence genes carried by each isolate. We performed a survival analysis to investigate the 28-day mortality predictors using the Cox regression model (primary endpoint), after evaluated the proportional hazard (PH) assumption. In case of non-PH assumption, the weighted Cox regression model was performed [54]. A bivariate analysis (univariate) was carried out and the variables for which the p-value was <0.10 in univariate analysis were included in the multivariate model. All variables entered as covariates were evaluated at the baseline. As secondary endpoint, a survival analysis was conducted using the Kaplan-Meier curves to assess the probability of death at 28 days. To facilitate the analysis, the resistome and virulome data obtained in sequencing were categorized as number of classes of virulence and resistance genes carried by each isolate. Because of the huge heterogeneity of treatment regimens among the included patients, antibiotics were coded according to the WHO ATC/DDD Index 2020 at the 4th level [55]. p values <0.05 were considered as statistically significant.

All statistical analyses were performed using R Statistical Software (version 4.0.2) [56].

4.5. Ethics

The source studies were approved by the Ethical Committees of the participating institutions, and all enrolled patients' legal guardians provided informed consent. Given the retrospective nature of the present study, ethical approval for this analysis was not necessary.

5. Conclusions

To conclude, this pilot study demonstrated the feasibility of investigating the association between neonatal sepsis mortality and the causative Enterobacterales isolates virulome. The limited sample size of our cohort did not allow us to determine the role of single virulence genes in neonatal GN-BSIs but grouping genes as the number of classes involved allowed us to investigate the impact of the entire virulome in neonatal sepsis outcome. This knowledge may be useful for predicting clinical outcomes, detecting virulent strains, and helping with vaccine development. Expanding research on anti-virulence molecules together with the development of new antibiotics could be crucial to improving the management of these fragile patients. Further research would be advisable to elucidate the correlation with the timing of sepsis onset to personalize the clinical approach. These findings need further exploration in larger global studies, ideally including host immunopathological response, in order to develop a tailor-made therapeutic strategy.

Supplementary Materials: The following are available online at https://www.mdpi.com/article/10.3390/antibiotics10060706/s1, Table S1: Isolate-specific accession numbers.

Author Contributions: Conceptualization, L.F., M.S.; methodology, D.D.C., F.C., A.P., Y.H.; validation, D.D.C., F.C., A.A.W., I.B., Y.H., K.L., I.M., J.B., A.A., G.V.Z., T.P., P.T.H., M.S.; formal analysis, L.F., D.D.C., F.C., A.P., A.A.W., K.L., I.M.; investigation, L.F., K.L., I.M.; resources, L.F., M.S.; data curation, L.F., Y.H.; writing—original draft preparation, L.F.; writing—review and editing, D.D.C., F.C., A.P., A.A.W., I.B., Y.H., K.L., I.M., J.B., A.A., G.V.Z., T.P., P.T.H., M.S.; funding acquisition, M.S. All authors have read and agreed to the published version of the manuscript.

Funding: This study and the APC were funded by Penta Foundation.

Institutional Review Board Statement: The source studies were approved by the Ethical Committees of the participating institutions. Given the retrospective nature of the present study, ethical approval for this analysis was not necessary.

Informed Consent Statement: All enrolled patients' legal guardians provided informed consent.

Data Availability Statement: The corresponding author confirms that he had full access to all the data in the study and had final responsibility for the decision to submit for publication. Whole-genome shotgun projects have been deposited in GenBank (project code PRJEB44870). The isolate-specific accession numbers are indicated in Supplementary Table S1.

Acknowledgments: We acknowledge "The NeoMero Consortium" for providing data regarding patients enrolled in the NeoMero 1 clinical trial for the present study.

Conflicts of Interest: The authors declare no conflict of interest related to this work.

References

1. Ballot, D.E.; Bandini, R.; Nana, T.; Bosman, N.; Thomas, T.; Davies, V.A.; Cooper, P.A.; Mer, M.; Lipman, J. A review of -multidrug-resistant Enterobacteriaceae in a neonatal unit in Johannesburg, South Africa. *BMC Pediatr.* **2019**, *19*, 320. [CrossRef]
2. Cole, B.K.; Ilikj, M.; McCloskey, C.B.; Chavez-Bueno, S. Antibiotic resistance and molecular characterization of bacteremia *Escherichia coli* isolates from newborns in the United States. *PLoS ONE* **2019**, *14*, e0219352. [CrossRef]
3. Li, S.; Liu, J.; Chen, F.; Cai, K.; Tan, J.; Xie, W.; Qian, R.; Liu, X.; Zhang, W.; Du, H.; et al. A risk score based on pediatric sequential organ failure assessment predicts 90-day mortality in children with Klebsiella pneumoniae bloodstream infection. *BMC Infect. Dis.* **2020**, *20*, 916. [CrossRef]
4. Nour, I.; Eldegla, H.E.; Nasef, N.; Shouman, B.; Abdel-Hady, H.; Shabaan, A.E. Risk factors and clinical outcomes for carbapenem-resistant Gram-negative late-onset sepsis in a neonatal intensive care unit. *J. Hosp. Infect.* **2017**, *97*, 52–58. [CrossRef]
5. Thatrimontrichai, A.; Premprat, N.; Janjindamai, W.; Dissaneevate, S.; Maneenil, G. Risk Factors for 30-Day Mortality in Neonatal Gram-Negative Bacilli Sepsis. *Am. J. Perinatol.* **2020**, *37*, 689–694. [CrossRef]

6. Sands, K.; Carvalho, M.J.; Portal, E.; Thomson, K.; Dyer, C.; Akpulu, C.; Andrews, R.; Ferreira, A.; Gillespie, D.; Hender, T.; et al. Characterization of antimicrobial-resistant Gram-negative bacteria that cause neonatal sepsis in seven low- and middle-income countries. *Nat. Microbiol.* **2021**, *6*, 512–523. [CrossRef]

7. Budhram, D.R.; Mac, S.; Bielecki, J.M.; Patel, S.N.; Sander, B. Health outcomes attributable to carbapenemase-producing Enterobacteriaceae infections: A systematic review and meta-analysis. *Infect. Control. Hosp. Epidemiol.* **2020**, *41*, 37–43. [CrossRef]

8. Burnham, J.P.; Lane, M.A.; Kollef, M.H. Impact of Sepsis Classification and Multidrug-Resistance Status on Outcome among Patients Treated with Appropriate Therapy. *Crit. Care Med.* **2015**, *43*, 1580–1586. [CrossRef] [PubMed]

9. Folgori, L.; Bernaschi, P.; Piga, S.; Carletti, M.; Cunha, F.P.; Lara, P.H.; de Castro Peixoto, N.C.; Alves Guimarães, B.G.; Sharland, M.; Araujo da Silva, A.R.; et al. Healthcare-Associated Infections in Pediatric and Neonatal Intensive Care Units: Impact of Underlying Risk Factors and Antimicrobial Resistance on 30-Day Case-Fatality in Italy and Brazil. *Infect. Control. Hosp. Epidemiol.* **2016**, *37*, 1302–1309. [CrossRef]

10. Gutiérrez-Gutiérrez, B.; Salamanca, E.; de Cueto, M.; Hsueh, P.R.; Viale, P.; Paño-Pardo, J.R.; Venditti, M.; Tumbarello, M.; Daikos, G.; Cantón, R.; et al. Effect of appropriate combination therapy on mortality of patients with bloodstream infections due to carbapenemase-producing Enterobacteriaceae (INCREMENT): A retrospective cohort study. *Lancet Infect. Dis.* **2017**, *17*, 726–734. [CrossRef]

11. Kyo, M.; Ohshimo, S.; Kosaka, T.; Fujita, N.; Shime, N. Impact of inappropriate empiric antimicrobial therapy on mortality in pediatric patients with bloodstream infection: A retrospective observational study. *J. Chemother.* **2019**, *31*, 388–393. [CrossRef] [PubMed]

12. Sakellariou, C.; Gürntke, S.; Steinmetz, I.; Kohler, C.; Pfeifer, Y.; Gastmeier, P.; Schwab, F.; Kola, A.; Deja, M.; Leistner, R. Sepsis Caused by Extended-Spectrum Beta-Lactamase (ESBL)-Positive, *K. pneumoniae* and *E. coli*: Comparison of Severity of Sepsis, Delay of Anti-Infective Therapy and ESBL Genotype. *PLoS ONE* **2016**, *11*, e0158039. [CrossRef]

13. Sianipar, O.; Asmara, W.; Dwiprahasto, I.; Mulyono, B. Mortality risk of bloodstream infection caused by either *Escherichia coli* or *Klebsiella pneumoniae* producing extended-spectrum β-lactamase: A prospective cohort study. *BMC Res. Notes* **2019**, *12*, 719. [CrossRef]

14. Tanır Basaranoglu, S.; Ozsurekci, Y.; Aykac, K.; Karadag Oncel, E.; Bıcakcigil, A.; Sancak, B.; Cengiz, A.B.; Kara, A.; Ceyhan, M. A comparison of blood stream infections with extended spectrum beta-lactamase-producing and non-producing *Klebsiella pneumoniae* in pediatric patients. *Ital. J. Pediatr.* **2017**, *43*, 79. [CrossRef]

15. Zarkotou, O.; Pournaras, S.; Tselioti, P.; Dragoumanos, V.; Pitiriga, V.; Ranellou, K.; Prekates, A.; Themeli-Digalaki, K.; Tsakris, A. Predictors of mortality in patients with bloodstream infections caused by KPC-producing *Klebsiella pneumoniae* and impact of appropriate antimicrobial treatment. *Clin. Microbiol. Infect.* **2011**, *17*, 1798–1803. [CrossRef]

16. Lisboa, T.; Waterer, G.; Rello, J. We should be measuring genomic bacterial load and virulence factors. *Crit. Care Med.* **2010**, *38* (Suppl. S10), S656–S662. [CrossRef]

17. Ron, E.Z. Distribution and evolution of virulence factors in septicemic Escherichia coli. *Int. J. Med. Microbiol.* **2010**, *300*, 367–370. [CrossRef]

18. Paczosa, M.K.; Mecsas, J. *Klebsiella pneumoniae*: Going on the Offense with a Strong Defense. *Microbiol. Mol. Biol. Rev.* **2016**, *80*, 629–661. [CrossRef]

19. Daga, A.P.; Koga, V.L.; Soncini, J.G.M.; de Matos, C.M.; Peruginim, M.R.E.; Pelisson, M.; Kobayashi, R.K.T.; Vespero, E.C. Escherichia coli Bloodstream Infections in Patients at a University Hospital: Virulence Factors and Clinical Characteristics. *Front. Cell Infect. Microbiol.* **2019**, *9*, 191. [CrossRef]

20. Du, F.; Wei, D.D.; Wan, L.G.; Cao, X.W.; Zhang, W.; Liu, Y. Evaluation of ompK36 allele groups on clinical characteristics and virulence features of *Klebsiella pneumoniae* from bacteremia. *J. Microbiol. Immunol. Infect.* **2019**, *52*, 779–787. [CrossRef]

21. Hung, W.T.; Cheng, M.F.; Tseng, F.C.; Chen, Y.S.; Shin-Jung Lee, S.; Chang, T.H.; Lin, H.H.; Hung, C.H.; Wang, J.L. Bloodstream infection with extended-spectrum beta-lactamase-producing *Escherichia coli*: The role of virulence genes. *J. Microbiol. Immunol. Infect.* **2019**, *52*, 947–955. [CrossRef] [PubMed]

22. Cheng, C.H.; Tsau, Y.K.; Kuo, C.Y.; Su, L.H.; Lin, T.Y. Comparison of extended virulence genotypes for bacteria isolated from pediatric patients with urosepsis, acute pyelonephritis, and acute lobar nephronia. *Pediatr. Infect. Dis. J.* **2010**, *29*, 736–740. [CrossRef]

23. Palma, N.; Gomes, C.; Riveros, M.; García, W.; Martínez-Puchol, S.; Ruiz-Roldán, L.; Mateu, J.; García, C.; Jacobs, J.; Ochoa, T.J.; et al. Virulence factors profiles and ESBL production in *Escherichia coli* causing bacteremia in Peruvian children. *Diagn. Microbiol. Infect. Dis.* **2016**, *86*, 70–75. [CrossRef]

24. Roy, S.; Datta, S.; DAS, P.; Gaind, R.; Pal, T.; Tapader, R.; Mukherjee, S.; Basu, S. Insight into neonatal septicaemic *Escherichia coli* from India with respect to phylogroups, serotypes, virulence, extended-spectrum-β-lactamases and association of ST131 clonal group. *Epidemiol. Infect.* **2015**, *143*, 3266–3276. [CrossRef] [PubMed]

25. Tapader, R.; Chatterjee, S.; Singh, A.K.; Dayma, P.; Haldar, S.; Pal, A.; Basu, S. The high prevalence of serine protease autotransporters of Enterobacteriaceae (SPATEs) in *Escherichia coli* causing neonatal septicemia. *Eur. J. Clin. Microbiol. Infect. Dis.* **2014**, *33*, 2015–2024. [CrossRef]

26. Burdet, C.; Clermont, O.; Bonacorsi, S.; Laouénan, C.; Bingen, E.; Aujard, Y.; Mentré, F.; Lefort, A.; Denamur, E. *Escherichia coli* bacteremia in children: Age and portal of entry are the main predictors of severity. *Pediatr. Infect. Dis. J.* **2014**, *33*, 872–879. [CrossRef]

27. Horváth, A.; Dobay, O.; Sahin-Tóth, J.; Juhász, E.; Pongrácz, J.; Iván, M.; Fazakas, E.; Kristóf, K. Characterisation of antibiotic resistance, virulence, clonality and mortality in MRSA and MSSA bloodstream infections at a tertiary-level hospital in Hungary: A 6-year retrospective study. *Ann. Clin. Microbiol. Antimicrob.* **2020**, *19*, 17. [CrossRef] [PubMed]

28. Zürn, K.; Lander, F.; Hufnagel, M.; Monecke, S.; Berner, R. Microarray Analysis of Group B Streptococci Causing Invasive Neonatal Early- and Late-onset Infection. *Pediatr. Infect. Dis. J.* **2020**, *39*, 449–453. [CrossRef]

29. Xu, M.; Fu, Y.; Kong, H.; Chen, X.; Chen, Y.; Li, L.; Yang, Q. Bloodstream infections caused by Klebsiella pneumoniae: Prevalence of bla(KPC), virulence factors and their impacts on clinical outcome. *BMC Infect. Dis.* **2018**, *18*, 358. [CrossRef]

30. Neonatal Infection Surveillance Database. Available online: https://www.neonin.org.uk/ (accessed on 15 April 2021).

31. Lutsar, I.; Chazallon, C.; Trafojer, U.; de Cabre, V.M.; Auriti, C.; Bertaina, C.; Calo Carducci, F.I.; Canpolat, F.E.; Esposito, S.; Fournier, I.; et al. Meropenem vs standard of care for treatment of neonatal late onset sepsis (NeoMero1): A randomised controlled trial. *PLoS ONE* **2020**, *15*, e0229380. [CrossRef]

32. CLAHRC-Infection Theme-Gram-Negative Project—Bloodstream Infection. Available online: http://www.clahrc-southlondon.nihr.ac.uk/infection/investigating-causes-and-care-septicaemia-caused-gram-negative-bacteria (accessed on 15 April 2021).

33. Derakhshan, S.; Najar Peerayeh, S.; Bakhshi, B. Association between Presence of Virulence Genes and Antibiotic Resistance in Clinical Klebsiella Pneumoniae Isolates. *Lab. Med.* **2016**, *47*, 306–311. [CrossRef] [PubMed]

34. Pop-Vicas, A.; Opal, S.M. The clinical impact of multidrug-resistant gram-negative bacilli in the management of septic shock. *Virulence* **2014**, *5*, 206–212. [CrossRef]

35. Viale, P.; Giannella, M.; Lewis, R.; Trecarichi, E.M.; Petrosillo, N.; Tumbarello, M. Predictors of mortality in multidrug-resistant Klebsiella pneumoniae bloodstream infections. *Expert Rev. Anti Infect. Ther.* **2013**, *11*, 1053–1063. [CrossRef]

36. Ray, S.; Anandm, D.; Purwar, S.; Samanta, A.; Upadhye, K.V.; Gupta, P.; Dhar, D. Association of high mortality with extended-spectrum β-lactamase (ESBL) positive cultures in community acquired infections. *J. Crit. Care* **2018**, *44*, 255–260. [CrossRef] [PubMed]

37. Mora-Rillo, M.; Fernández-Romero, N.; Navarro-San Francisco, C.; Díez-Sebastián, J.; Romero-Gómez, M.P.; Fernández, F.A.; López, J.R.A.; Mingorance, J. Impact of virulence genes on sepsis severity and survival in Escherichia coli bacteremia. *Virulence* **2015**, *6*, 93–100. [CrossRef]

38. Kim, D.; Park, B.Y.; Choi, M.H.; Yoon, E.J.; Lee, H.; Lee, K.J.; Park, Y.S.; Shin, J.H.; Uh, Y.; Shin, K.S.; et al. Antimicrobial resistance and virulence factors of Klebsiella pneumoniae affecting 30 day mortality in patients with bloodstream infection. *J. Antimicrob. Chemother.* **2019**, *74*, 190–199. [CrossRef]

39. Namikawa, H.; Niki, M.; Niki, M.; Yamada, K.; Nakaie, K.; Sakiyama, A.; Oinuma, K.I.; Tsubouchi, T.; Tochino, Y.; Takemoto, Y.; et al. Clinical and virulence factors related to the 30-day mortality of Klebsiella pneumoniae bacteremia at a tertiary hospital: A case-control study. *Eur. J. Clin. Microbiol. Infect. Dis.* **2019**, *38*, 2291–2297. [CrossRef]

40. Geisinger, E.; Isberg, R.R. Interplay between Antibiotic Resistance and Virulence during Disease Promoted by Multidrug-Resistant Bacteria. *J. Infect. Dis.* **2017**, *215* (Suppl. S1), S9–S17. [CrossRef]

41. Gharrah, M.M.; El-Mahdy, A.M.; Barwa, R.F. Association between Virulence Factors and Extended Spectrum Beta-Lactamase Producing Klebsiella pneumoniae Compared to Nonproducing Isolates. *Interdiscip. Perspect. Infect. Dis.* **2017**, *2017*, 7279830. [CrossRef]

42. Cepas, V.; Soto, S.M. Relationship between Virulence and Resistance among Gram-Negative Bacteria. *Antibiotics* **2020**, *9*, 719. [CrossRef]

43. Gomez-Simmonds, A.; Uhlemann, A.C. Clinical Implications of Genomic Adaptation and Evolution of Carbapenem-Resistant *Klebsiella pneumoniae. J. Infect. Dis.* **2017**, *215* (Suppl. S1), S18–S27. [CrossRef]

44. Abd El-Baky, R.M.; Ibrahim, R.A.; Mohamed, D.S.; Ahmed, E.F.; Hashem, Z.S. Prevalence of Virulence Genes and Their Association with Antimicrobial Resistance Among Pathogenic, *E. coli* Isolated from Egyptian Patients with Different Clinical Infections. *Infect. Drug Resist.* **2020**, *13*, 1221–1236. [CrossRef]

45. Beceiro, A.; Tomás, M.; Bou, G. Antimicrobial resistance and virulence: A successful or deleterious association in the bacterial world? *Clin. Microbiol. Rev.* **2013**, *26*, 185–230. [CrossRef] [PubMed]

46. Høiby, N.; Bjarnsholt, T.; Givskov, M.; Molin, S.; Ciofu, O. Antibiotic resistance of bacterial biofilms. *Int. J. Antimicrob. Agents.* **2010**, *35*, 322–332. [CrossRef]

47. Johnson, T.J.; Nolan, L.K. Pathogenomics of the virulence plasmids of *Escherichia coli. Microbiol. Mol. Biol. Rev.* **2009**, *73*, 750–774. [CrossRef]

48. Vergnano, S.; Menson, E.; Kennea, N.; Embleton, N.; Russell, A.B.; Watts, T.; Robinson, M.J.; Collinson, A. Heath PT. Neonatal infections in England: The NeonIN surveillance network. *Arch. Dis. Child. Fetal. Neonatal. Ed.* **2011**, *96*, F9–F14. [CrossRef]

49. Magiorakos, A.P.; Srinivasan, A.; Carey, R.B.; Carmeli, Y.; Falagas, M.E.; Giske, C.G.; Harbarth, S.; Hindler, J.F.; Kahlmeter, G.; Olsson-Liljequist, B.; et al. Multidrug-resistant, extensively drug-resistant and pandrug-resistant bacteria: An international expert proposal for interim standard definitions for acquired resistance. *Clin. Microbiol. Infect.* **2012**, *18*, 268–281. [CrossRef] [PubMed]

50. Bankevich, A.; Nurk, S.; Antipov, D.; Gurevich, A.A.; Dvorkin, M.; Kulikov, A.S.; Lesin, V.M.; Nikolenko, S.I.; Pham, S.; Prjibelski, A.D.; et al. A new genome assembly algorithm and its applications to single-cell sequencing. *J. Comput. Biol.* **2012**, *19*, 455–477. [CrossRef] [PubMed]

51. Jia, B.; Raphenya, A.R.; Alcock, B.; Waglechner, N.; Guo, P.; Tsang, K.K.; Lago, B.A.; Dave, B.M.; Pereira, S.; Sharma, A.N.; et al. CARD 2017: Expansion and model-centric curation of the comprehensive antibiotic resistance database. *Nucleic Acids Res.* **2017**, *45*, D566–D573. [CrossRef]
52. Zankari, E.; Hasman, H.; Cosentino, S.; Vestergaard, M.; Rasmussen, S.; Lund, O.; Aarestrup, F.M.; Larsen, M.V. Identification of acquired antimicrobial resistance genes. *J. Antimicrob. Chemother.* **2012**, *67*, 2640–2644. [CrossRef]
53. Chen, L.; Zheng, D.; Liu, B.; Yang, J.; Jin, Q. VFDB 2016: Hierarchical and refined dataset for big data analysis—10 years on. *Nucleic Acids Res.* **2016**, *44*, D694–D697. [CrossRef] [PubMed]
54. Schemper, M.; Wakounig, S.; Heinze, G. The estimation of average hazard ratios by weighted Cox regression. *Stat. Med.* **2009**, *28*, 2473–2489. [CrossRef] [PubMed]
55. ATC/DDD Index 2020. Available online: https://www.whocc.no/atc_ddd_index/ (accessed on 23 November 2020).
56. R Core Team. *R: A Language and Environment for Statistical Computing*; R Foundation for Statistical Computing: Vienna, Austria, 2020; Available online: https://www.R-project.org/ (accessed on 15 April 2021).

Article

Influence of Nisin-Biogel at Subinhibitory Concentrations on Virulence Expression in *Staphylococcus aureus* Isolates from Diabetic Foot Infections

Carolina Jesus, Rui Soares, Eva Cunha, Miguel Grilo, Luís Tavares and Manuela Oliveira *

CIISA—Centro de Investigação Interdisciplinar em Sanidade Animal, Faculdade de Medicina Veterinária, Universidade de Lisboa, Avenida da Universidade Técnica, 1300-477 Lisboa, Portugal; carolinaojesus@gmail.com (C.J.); rmsoares@fmv.ulisboa.pt (R.S.); evacunha@fmv.ulisboa.pt (E.C.); miguelgrilo@fmv.ulisboa.pt (M.G.); ltavares@fmv.ulisboa.pt (L.T.)
* Correspondence: moliveira@fmv.ulisboa.pt

Abstract: A new approach to diabetic foot infections (DFIs) has been investigated, using a nisin-biogel combining the antimicrobial peptide (AMP) nisin with the natural polysaccharide guar-gum. Since in in vivo conditions bacteria may be exposed to decreased antimicrobial concentrations, known as subinhibitory concentrations (sub-MICs), effects of nisin-biogel sub-MIC values corresponding to 1/2, 1/4 and 1/8 of nisin's minimum inhibitory concentration (MIC) on virulence expression by six *Staphylococcus aureus* DFI isolates was evaluated by determining bacteria growth rate; expression of genes encoding for staphylococcal protein A (*spA*), coagulase (*coa*), clumping factor A (*clfA*), autolysin (*atl*), intracellular adhesin A (*icaA*), intracellular adhesin D (*icaD*), and the accessory gene regulator I (*agrI*); biofilm formation; Coa production; and SpA release. Nisin-biogel sub-MICs decreased bacterial growth in a strain- and dose-dependent manner, decreased *agrI*, *atl* and *clfA* expression, and increased *spA*, *coa*, *icaA* and *icaD* expression. Biofilm formation increased in the presence of nisin-biogel at 1/4 and 1/8 MIC, whereas 1/2 MIC had no effect. Finally, nisin-biogel at sub-MICs did not affect coagulase production, but decreased SpA production in a dose-dependent manner. Results highlight the importance of optimizing nisin-biogel doses before proceeding to in vivo trials, to reduce the risk of virulence factor's up-regulation due to the presence of inappropriate antimicrobial concentrations.

Keywords: diabetic foot infections; *Staphylococcus aureus*; subinhibitory concentrations; virulence-related genes; biofilm

---

Citation: Jesus, C.; Soares, R.; Cunha, E.; Grilo, M.; Tavares, L.; Oliveira, M. Influence of Nisin-Biogel at Subinhibitory Concentrations on Virulence Expression in *Staphylococcus aureus* Isolates from Diabetic Foot Infections. *Antibiotics* **2021**, *10*, 1501. https://doi.org/10.3390/antibiotics10121501

Academic Editor: Ágnes Pál-Sonnevend

Received: 7 November 2021
Accepted: 3 December 2021
Published: 7 December 2021

Publisher's Note: MDPI stays neutral with regard to jurisdictional claims in published maps and institutional affiliations.

1. Introduction

Diabetes *mellitus* (DM) is a lifelong metabolic disorder that affects approximately 537 million people worldwide [1]. The development of diabetic foot ulcers (DFUs) is a serious complication associated with the DM triad of neuropathy, vasculopathy and immunopathy [2]. The severe loss of skin protective barriers creates a chance for tissue colonization by opportunistic microorganisms, including *Staphylococcus aureus*. Diabetic foot infections (DFIs) usually become chronic and result in increased patient morbidity and mortality, the most common DM complication requiring hospitalization and often resulting in lower-extremity amputations [3].

S. aureus is a Gram-positive bacterium that expresses several regulatory and virulence factors, which contributes to its success as a human opportunistic pathogen [4]. In the first step of a staphylococcal infection, the adhesion phase, the production of several cell surface-associated factors occurs, including clumping factor A (ClfA), staphylococcal protein A (SpA), and coagulase (Coa) that facilitate tissue attachment and evasion of the host immune system [4,5]. Staphylococcal major autolysin (Atl) is also relevant for staphylococcal attachment to surfaces, also participating in bacterial cell wall degradation, lysis mediated biofilm development and secretion of cytoplasmic proteins [6–8].

The ability to form biofilms is considered a major staphylococcal pathogenic feature [9]. Biofilms are characterized by the growth of adherent bacterial populations inside a self-produced matrix of extracellular polymeric substances, conferring this sessile mode of life survival advantages, including enhanced antimicrobial resistance. The intercellular adhesion (*ica*) locus, *icaADBC*, is associated with cell-to-cell adhesion, being responsible for the biosynthesis of the biofilm exopolysaccharide intercellular adhesion (PIA) [10–13]. The coordination between *S. aureus* adhesion and detachment is regulated by Quorum Sensing (QS). The accessory gene regulator (*agr*) of *S. aureus* is a QS global transcriptional regulator, controlling virulence factors and biofilm expression in a time and population density-dependent manner [10,14,15].

Another key factor for the success of *S. aureus* as a human pathogen is the ability to rapidly develop or acquire multiple antibiotic resistance determinants. Nowadays, the development and spread of pathogenic bacteria that are resistant to conventional antibiotics is a major public health concern, which makes it urgent to discover, develop, and implement new effective therapeutic strategies [16–20]. Antimicrobial peptides (AMPs) have been proposed as promising therapeutic candidates to inhibit bacterial growth which can be used synergistically with antibiotics [17]. One of the most studied AMPs is nisin, a lantibiotic produced by *Lactococcus lactis* which interacts with the bacterial cell wall precursor lipid II, inhibiting cell wall synthesis, and uses lipid II as a docking molecule for subsequent pore formation [21]. Nisin was first commercialized as a food preservative, being recognized as safe by the Food and Agriculture Organization/World Health Organization (FAO/WHO) [17,22,23]. In recent years, it started to be investigated in veterinary and pharmaceutical fields, including for the management of bacterial infections, as this polypeptide presents antimicrobial activity against a broad spectrum of Gram-positive bacteria [17]. *Staphylococcus* species have a remarkable susceptibility to nisin, which represent an advantage for nisin application towards skin infections, including DFIs, and the treatment of multiple drug resistant systemic infections [17,22].

As AMPs can be inactivated or degraded before achieving their target, different methods for AMPs delivery have been widely investigated, aiming at increasing their clinically efficacy. Natural polysaccharides have been considered promising drug delivery systems mainly due to their non-toxicity, sustainability, biodegradability, biocompatibility, abundance, availability, and cost-effectiveness [24]. Guar gum is a natural, uncharged, and water-soluble polysaccharide that has been largely used for targeted drug delivery, promoting a controlled drug release and availability [25,26]. Considering all the promising features of nisin and guar gum, a guar gum gel-based delivery system for nisin, nisin-biogel, has been developed by our research team as an alternative or complementary strategy to conventional antibiotics used in DFIs treatment [27]. Previous studies have shown that the nisin-biogel had inhibitory capacity towards *S. aureus* DFI isolates either in their planktonic and biofilm forms [27], could be applied in combination with conventional antibiotics and antiseptics to improve their efficacy [28,29], maintained its activity when stored at temperatures below 22 °C for 24 months [30], exhibited no significant levels of cytotoxicity on human keratinocyte cells [30] and was able to diffuse and keep its antimicrobial activity in a DFU collagen three-dimensional (3D) model established to mimic the DFU environment [29]. However, before proceeding to in vivo trials, the effect of drug therapeutic doses on virulence determinants expression by *S. aureus* must be addressed, as it may affect infection pathogenesis. Minimum inhibitory concentration (MIC) is defined as the lowest concentration of an antimicrobial that inhibits the growth of most of the target bacterial population, under controlled in vitro conditions [31]. However, in DFIs patients there is a poor diffusion of the antimicrobials that results from DM patient's compromised blood circulation and low perfusion due to angiopathy [31,32]. As such, in in vivo infections, bacteria may be exposed to a reduced effective concentration of antimicrobials, referred to as subinhibitory concentrations (sub-MICs), which can lead to a wide variety of physiological and morphological effects on bacteria and, consequently, affect infection pathogenesis [31,33]. Besides promoting bacterial resistance, sub-MICs of antimicrobials can modulate the virulence of *S. aureus*, by influencing gene expression, biofilm production and the QS system, which may impact the outcome of staphylococcal infections [31,33–36]. As such, the present study evaluated the effects of nisin-biogel sub-MICs on *S. aureus* DFI isolates growth rate, virulence-related genes expression, namely of *agrI*, *spA*, *clfA*, *coa*, *atl*, *icaA* and *icaD*, ability to form biofilm, Coa production and SpA release, aiming at confirming its suitability for in vivo administration.

2. Results

2.1. Effect of Nisin-Biogel Sub-MICs on S. aureus DFI Isolates Growth Rate

The effect of nisin-biogel sub-MICs on *S. aureus* DFI isolates A 5.2, A 6.3, B 1.1, B 14.2, Z 1.1, and Z 5.2 growth rate is represented in Figure 1. Both in the presence or absence of nisin-biogel at sub-MICs, all *S. aureus* clinical isolates growth curves presented the typical sigmoidal pattern with the three phases of bacterial growth curves—the lag phase, the exponential phase, and the stationary phase. Nevertheless, different clinical isolates showed different growth rates, with isolate A 6.3 having the highest growth rate, and isolate A 5.2 having the lowest one. Nisin-biogel at sub-MICs slowed bacterial growth, delaying the beginning of the exponential growth phase. Nisin-biogel concentration equivalent to 1/8 MIC was the one that promoted the lower change in bacterial growth, while the 1/2 MIC was the concentration that promoted a higher change on the bacterial growth. For some *S. aureus* clinical isolates, including isolates Z 5.2, A 6.3, B 1.1 and Z 1.1, there was an increase in bacterial growth in the stationary phase in the presence of nisin-biogel at sub-MICs.

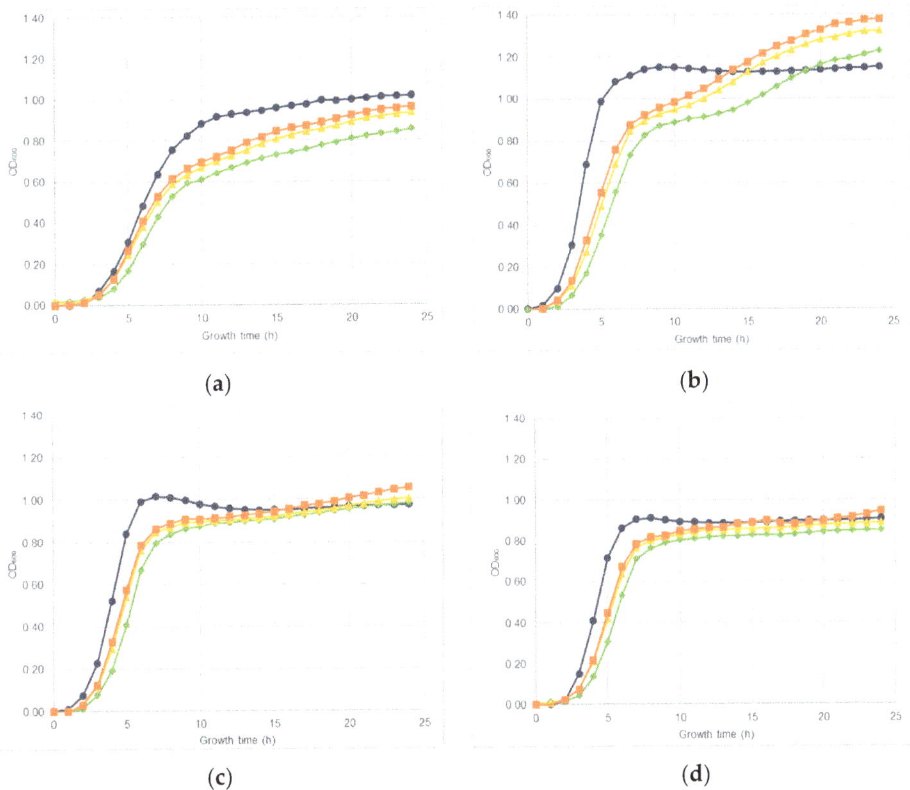

(a)

(b)

(c)

(d)

Figure 1. *Cont.*

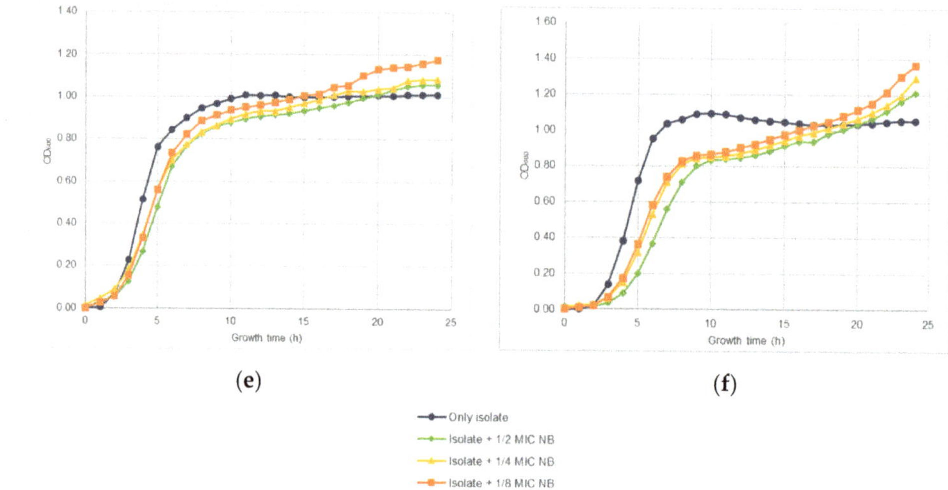

Figure 1. Growth curves obtained for *S. aureus* clinical isolates during 24 h at 37 °C with shaking (150 rpm) in the absence or in the presence of different sub-MICs of nisin-biogel. Results are presented as mean values of three independent assays, reflecting bacterial growth for each *S. aureus* clinical isolate under the different conditions tested. (**a**): Strain ID A 5.2, isolated from a DFI aspirate (*agrI* +; *spA* +; *atl* +; *clfA* +; *coa* +; *icaA* +; *icaD* +; Biofilm production +; MSSA). (**b**): Strain ID A 6.3, isolated from a DFI aspirate (*agrI* +; *spA* +; *atl* +; *clfA* +; *coa* +; *icaA* +; *icaD* +; Biofilm production +; MSSA). (**c**): Strain ID B 1.1, isolated from a DFI biopsy (*agrI* +; *spA* +; *atl* +; *clfA* +; *coa* +; *icaA* +; *icaD* +; Biofilm production +; MRSA). (**d**): Strain ID B 14.2, isolated from a DFI biopsy (*agrI* +; *spA* +; *atl* +; *clfA* +; *coa* +; *icaA* +; *icaD* +; Biofilm production +; MRSA). (**e**): Strain ID Z 1.1, isolated from a DFI swab (*agrI* +; *spA* +; *atl* +; *clfA* +; *coa* +; *icaA* +; *icaD* +; Biofilm production +; MRSA). (**f**): Strain ID Z 5.2, isolated from a DFI swab (*agrI* +; *spA* +; *atl* +; *clfA* +; *coa* +; *icaA* +; *icaD* +; Biofilm production +; MSSA). MIC: minimum inhibitory concentration. NB: nisin-biogel. OD_{600}: optical density at 600 nm.

2.2. S. aureus DFI Isolates Gene Expression Kinetics

S. aureus clinical isolates virulence genes expression kinetics was accessed using RT-qPCR assays. Figure 2a,b show that for *agrI* and *atl*, the highest expression levels were reached at 6 hours' incubation, while for *spA* and *clfA* it was at 4 h. For *coa*, maximum expression levels were reached at 3 h. Although the optical expression of the different genes occurred at different periods of bacterial growth, all revealed a considerable expression at 4 h' incubation, which allowed to select this time point as the more adequate for further evaluation of the effects of nisin-biogel sub-MICs on *agrI*, *spA*, *clfA*, *atl* and *coa* expression. Genes *icaA* and *icaD* reach their maximum expression levels at 48 h (Figure 3), which allowed to select this time point to *icaA* and *icaD* expression assays.

2.3. Effect of Nisin-Biogel at Sub-MICs on Gene Expression by S. aureus DFI Isolates

The effect of nisin-biogel and clindamycin at sub-MICs on virulence genes expression by *S. aureus* DFI isolates was investigated using RT-qPCR assays. As shown in Figure 4 and Table S1 (available as Supplementary Data), the effects depended on the virulence factor, on the *S. aureus* clinical isolate, and on the subinhibitory concentration under study. Overall, nisin-biogel and clindamycin at sub-MICs values significantly decreased the expression of *agrI*, with nisin-biogel at 1/2 MIC being the one that reduced the expression of *agrI* the most, and 1/8 MIC the one that least reduce *agrI* expression (Figure 4a). Nisin-biogel at sub-MICs increased the expression of *spA*, while clindamycin at 1/2 MIC significantly decreased the expression of this gene (Figure 4b). For *atl*, nisin-biogel at sub-MICs significantly decreased its expression, with 1/2 MIC being the one that least decreased *atl* expression. Unlike nisin-biogel at sub-MICs, clindamycin at 1/2 MIC increased the expression of this gene (Figure 4c). Nisin-biogel at sub-MICs significantly decreased *clfA* expression, with 1/2 MIC being responsible for the highest increase in the expression of this gene. On the other hand, clindamycin at 1/2 MIC significantly increased *clfA* expression (Figure 4d). For *coa*, both nisin-biogel and clindamycin at sub-MICs increased gene expression and, regarding nisin-biogel, 1/2 MIC was the one that increased *coa* expression the most (Figure 4e). For *icaA*, nisin-biogel at 1/4 and 1/8 MIC slightly increased the expression of this gene, while at 1/2 MIC slightly decreased *icaA* expression. Clindamycin at 1/2 MIC decreased *icaA* expression (Figure 4f). Finally, for *icaD*, nisin-biogel sub-MICs slightly increased gene expression, while clindamycin at 1/2 MIC decreased its expression (Figure 4g).

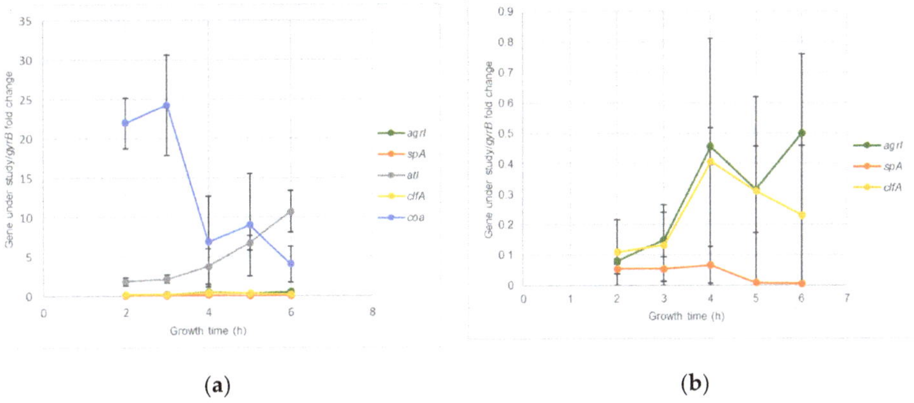

(a)   (b)

Figure 2. *agrI, spA, clfA, atl* and *coa* expression kinetics by *S. aureus* during a five hour growth period. Results are expressed as 'gene under study/*gyrB*' fold changes at 2, 3, 4, 5 and 6 h. Values are presented as means values ± SD (isolates A 5.2 and Z 5.2). (**b**) corresponds to an amplification of part of (**a**), so the y axis has different scales in the two figures. *agrI*: accessory gene regulator I; *spA*: gene encoding staphylococcal protein A; *atl*: gene encoding autolysin; *clfA*: gene encoding clumping factor A; *coa*: gene encoding coagulase; *gyrB*: gene encoding gyrase B.

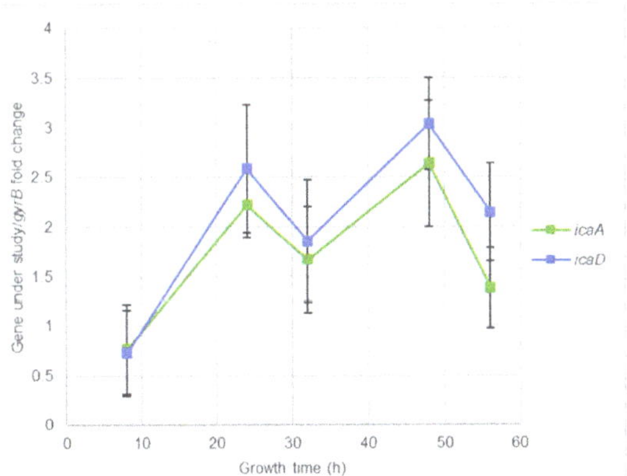

Figure 3. *icaA* and *icaD* expression kinetics for 48 h. Results are expressed as 'gene under study/*gyrB*' fold changes at 8, 24, 32, 48, and 56 h. Values are presented as means values ± SD (isolates A 5.2 and Z 5.2). *icaA*: gene encoding intracellular adhesin A; *icaD*: gene encoding intracellular adhesin D; *gyrB*: gene encoding gyrase B.

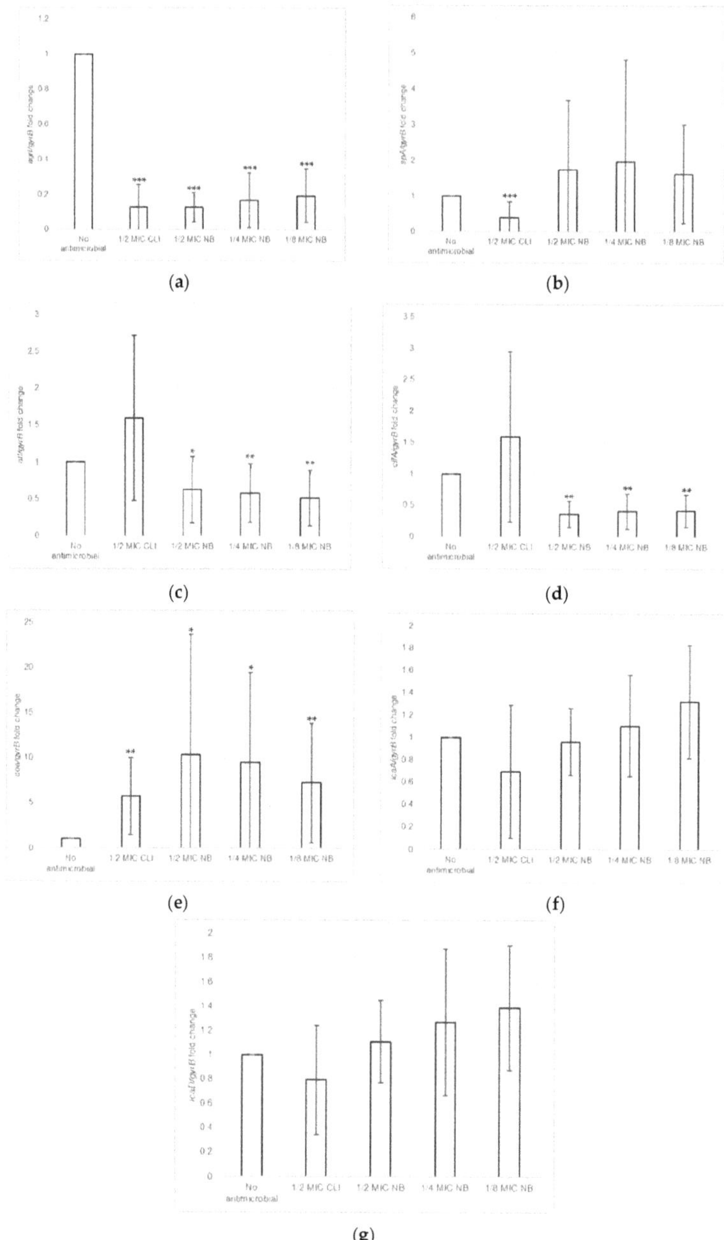

Figure 4. (**a–g**): Effects of nisin-biogel at 1/2 MIC, 1/4 MIC and 1/8 MIC and clindamycin at 1/2 MIC on *agrI*, *spA*, *atl*, *clfA*, *coa*, *icaA* and *icaD* mRNA expression, respectively. Results are expressed as n-fold differences in the 'gene under study/*gyrB*' ratio in the presence of the different conditions described above relative to 'gene under study/*gyrB*' ration in the growth control (no antimicrobial). Values are means ± SD (two repeated different experiments). Asterisks indicate statistically significant differences between treatments and control (* = $p < 0.05$; ** = $p < 0.01$; *** = $p < 0.001$). NB: nisin-biogel; CLI: clindamycin; MIC: minimum inhibitory concentration. *agrI*: accessory gene regulator I; *spA*: gene encoding staphylococcal protein A; *atl*: gene encoding autolysin; *clfA*: gene encoding clumping factor A; *coa*: gene encoding coagulase; *icaA*: gene encoding intracellular adhesin A; *icaD*: gene encoding intracellular adhesin D; *gyrB*: gene encoding gyrase B.

2.4. Effect of Nisin-Biogel Sub-MICs on the Ability of S. aureus DFI Isolates to Form Biofilm

Biofilm formation by *S. aureus* DFI isolates in the presence of nisin-biogel and clindamycin at sub-MICs was determined using a microtiter assay. Results shown in Table S2 (available as Supplementary Data) demonstrate that different *S. aureus* clinical isolates have different responses to nisin-biogel and clindamycin at sub-MICs. In Figure 5 it is possible to observe that, overall, nisin-biogel at 1/4 and clindamycin at 1/2 MIC exhibit a trend to increase the ability of *S. aureus* isolates to form biofilm, and, in the presence of nisin-biogel at 1/8 MIC, biofilm formation significantly increases. Oppositely, nisin-biogel at 1/2 MIC had no effect in the ability of *S. aureus* clinical isolates to form biofilm.

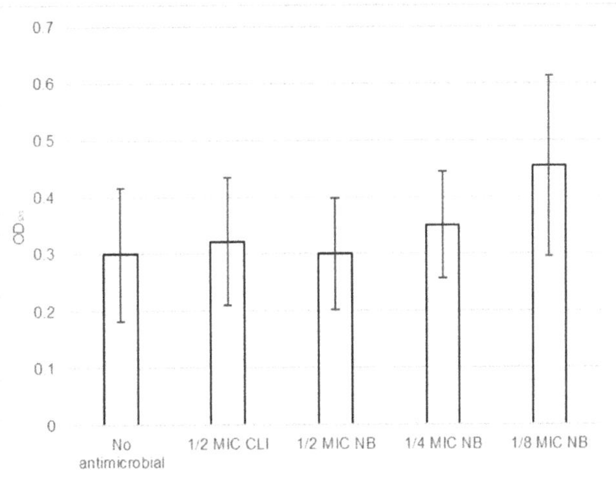

Figure 5. Effects of nisin-biogel at 1/2 MIC, 1/4 MIC and 1/8 MIC and clindamycin at 1/2 MIC on the ability of *S. aureus* DFI isolates to form biofilm. Values are presented as means $\pm$ SD (three repeated different experiments). Asterisks indicate statistically significant differences between treatments and control. NB: nisin-biogel; CLI: clindamycin; MIC: minimum inhibitory concentration. OD$_{570}$: optical density at 570 nm.

2.5. Effect of Nisin-Biogel Sub-MICs on Coa Production by S. aureus DFI Isolates

Coagulase test was used to monitor coagulase production by *S. aureus* clinical isolates in presence and absence of nisin-biogel and clindamycin at sub-MICs. Results were considered valid if the control plasma showed no signs of clotting in all assays.

Different clinical isolates showed different coagulase production ability over 4 h and after 24 h of incubation in the presence or absence of nisin-biogel and clindamycin at sub-MICs, as shown in Tables S3 and S4 (available as Supplementary Data). Besides depending on the *S. aureus* clinical isolate, the effects of antimicrobials sub-MICs on coagulation varied according to the bacterial growth period. During the first 4 h, in which results were monitored on an hourly basis, clotting in the absence of nisin-biogel was higher than or equal to the one obtained in the presence of nisin-biogel at sub-MICs. After 24 h incubation, most clinical isolates showed the same degree of clotting in the presence or absence of nisin-biogel. Regarding the effect of clindamycin at 1/2 MIC, for isolates A 5.2, A 6.3, and Z 5.2 there was no signs of coagulation in the presence of the antibiotic, whereas for isolates B 1.1, B 14.2 and Z 1.1, clotting degree was equal or higher than the one obtained in the absence of clindamycin at 1/2 MIC.

2.6. Effect of Nisin-Biogel Sub-MICs on SpA Release by S. aureus DFI Isolates

The effect of nisin-biogel and clindamycin at sub-MICs on SpA release by *S. aureus* DFI isolates was investigated using Protein A ELISA Kit. Results shown in Table S5 (available as Supplementary Data) demonstrate that different *S. aureus* clinical isolates have different responses regarding SpA release in the presence of antimicrobials sub-MICs. Figure 6 shows the overall effects of nisin-biogel and clindamycin at sub-MICs on the release of SpA by *S. aureus* DFI isolates. Nisin-biogel at 1/2 significantly decreased the release of SpA, and the other conditions under study exhibited a trend to decrease SpA release by *S. aureus* clinical isolates.

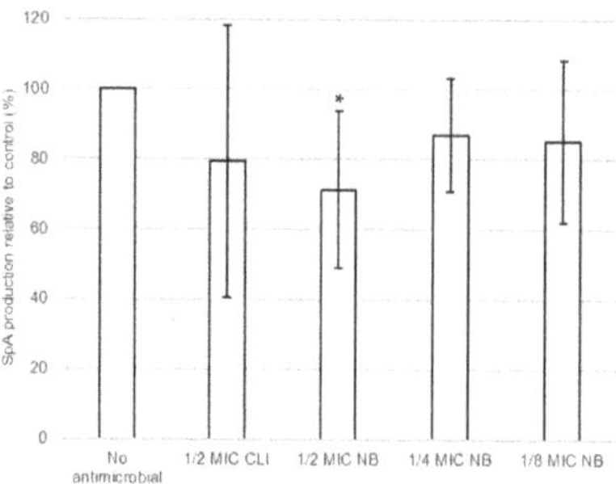

Figure 6. Effects of nisin-biogel at 1/2 MIC, 1/4 MIC and 1/8 MIC and clindamycin at 1/2 MIC on SpA production by *S. aureus* DFI isolates. Results are the ratios of the amount of SpA (pg/mL) in the bacterial supernatants incubated with nisin-biogel and clindamycin to the mean amount of SpA (pg/mL) incubated without antimicrobials and are expressed as supernatants. Values are presented as mean values ± SD. Asterisks indicate statistically significant differences between treatments and control (* = $p < 0.05$). NB: nisin-biogel; CLI: clindamycin; MIC: minimum inhibitory concentration. SpA: staphylococcal protein A.

3. Discussion

Diabetes *mellitus* (DM) is one of the most disseminated chronic diseases worldwide. The last International Diabetes Federation (IDF) Atlas states that, in 2021, 537 million adults aged over 20 years were living with diabetes worldwide. In Portugal, there were over 990 million cases of diabetes registered this year, with the cost associated to each diabetic patient being estimated to be approximately 2.990 euros (https://diabetesatlas.org/data/en/country/159/pt.html, accessed on 27 November 2021), proving that diabetes represents a significant health burden in our country [37].

A frequent and most devastating consequence of diabetes is the development of diabetic foot ulcers (DFU), which in about 50% of the cases become infected, developing to diabetic foot infections (DFI) [38]. Besides the conventional therapeutic protocols available for DFI treatment (e.g., debridement, wound healing agents, surgery and antibiotic therapy), several advanced therapeutics are also available, including bacteriophage therapy, negative-pressure wound therapy, hyperbaric oxygen therapy, stem cell therapy and off-loading [38]. Antimicrobial peptides are also a promising approach for DFI treatment; however, as observed for conventional antibiotics, angiopathy and low perfusion associated with DFU pathogenesis may prevent antimicrobials to reach the infected DFUs at effective concentration, leading to the presence of sub-MICs at the site of infection, which may impair treatment success [31,32].

The presence of antimicrobials sub-MICs can have several effects on bacteria, including the inhibition of bacterial growth [39,40] and therefore on infection progression [35,39]. As such, the effects of nisin-biogel at sub-MICs on bacterial growth were assessed. Previously, Field et al. [17] showed that nisin at sublethal concentrations slightly increased the lag period of *S. aureus* growth curve, which supports the results from this study, as the nisin-biogel sub-MICs values tested caused a reduction on *S. aureus* DFI isolates growth by increasing the lag phase. This reduction occurred in a dose-dependent manner, with the nisin-biogel at 1/2 MIC being the concentration that decreased bacterial growth the most, whereas 1/8 MIC was the one that least affected bacterial growth. In the stationary growth phase, the *S. aureus* clinical isolates Z 5.2, A 6.3, B 1.1 and Z 1.1 seem capable of adapting to the presence of nisin-biogel at sub-MICs, presenting an increase in growth rate.

S. aureus expresses a multitude of virulence factors in a coordinated manner, and many of them are under the control of the *agr* quorum sensing system. The *agr* gene positively controls the expression of many exotoxins, mainly produced after the end of the exponential growth phase, allowing bacteria to spread from the colonization sites to deeper tissues, and negatively controls the transcription of some cell wall-associated proteins, mainly synthesized during exponential growth [41]. According to our results, *agrI* expression levels by the *S. aureus* isolates tested only started to be relevant after 4 h, a period during which bacterial adherence to the host tissues already occurred in vivo. At this time point, the cell-surface proteins start to be down-regulated, whereas the virulence genes associated with bacterial dissemination and biofilm formation start to be up-regulated [4,42,43]. Peng et al. [44] showed that *agr* activity is required for post-exponential phase expression of various secreted proteins, which supports the fact that in the present study *agr* mRNA expression has its peak at 6 h.

SpA is a microbial surface protein that plays an important role in interfering with host defenses, inhibiting antibody-mediated phagocytosis. Staphylococcal extracellular protein, Coa, also contributes for the persistence of *S. aureus* in the host cell, by stimulating clotting formation in plasma, inhibiting host clearance mechanisms [45,46]. Both proteins are mainly expressed in the early stage of *S. aureus* growth, i.e., before 3–4 h. A study by Vandenesch et al. [47] suggested that *spA* mRNA is synthetized for a brief period in the beginning of the exponential phase and then is switched off. Moreover, according to Lebeau et al. [48], *coa* mRNA is mainly expressed at the early stages of *S. aureus* growth. Results from this study also showied that *coa* and *spA* mRNA are mostly expressed at the early exponential growth phase.

ClfA is also a staphylococcal surface protein that binds to Fg, allowing bacteria to colonize traumatized tissue and, later, form biofilms [49]. *clfA* is mainly expressed after 4 h of bacterial growth, i.e., at the late exponential growth phase. Results by Josefsson et al. [50] show that *clfA* expression is higher at 6 h and remains high until 24 h of *S. aureus* growth, which is not entirely in line with what we obtained in the present study, suggesting that different *S. aureus* strains may have different *clfA* mRNA expression kinetics. Atl protein has a multitude of functions, including attachment, bacterial cell wall degradation, biofilm development and bacteriolytic activity [51]. In the present study, *atl* expression increased with *S. aureus* growth period, within a time frame of 6 h, probably due to the increased growth rate over this time period. These results are in accordance with those by Oshida et al. [52] that also reported an increase in *atl* expression during the exponential growth phase.

One of the most relevant staphylococcal virulence factors is biofilm production, conferring bacteria a wide range of adaptive advantages that contribute to bacteria survival and persistence in the host [32]. Thus, *icaA* and *icaD*, involved in *icaADBC*-dependent biofilm formation, are mainly expressed latter during *S. aureus* growth. According to Atshan et al., 2013, *icaA* and *icaD* are up-regulated at 24 h of *S. aureus* growth and not at 48 h as observed in the present study, which suggests that *ica* mRNA expression kinetics may differ between *S. aureus* strains [53]. Furthermore, Patel et al. [54] evaluated gene expression during *Staphylococcus epidermidis* biofilm formation and, in accordance with the present study, the expression levels of *icaA* and *icaD* at 48 h were the highest ones. Although this study focused on *S. aureus*, *S. epidermidis* also produces biofilm in an *icaADBC*-dependent manner, allowing to suggest that the time frame required for the expression of *icaA* and *icaD* may be identical in both species.

Several studies have shown that the use of antimicrobials at sub-MICs modulates virulence gene expression and, consequently, influences bacterial pathogenicity [21,39]. Determining the effects of antimicrobials sub-MICs on virulence genes expression may provide important information for the rational use of antimicrobials in clinical practice [39]. One of the antibiotics currently applied as a DFIs alternative treatment is clindamycin, used in this study as a control to compare the effects of nisin-biogel at sub-MICs on *S. aureus* gene expression [21,55].

Subinhibitory levels of nisin-biogel decreased the expression of the regulatory gene *agrI* in a dose-dependent manner, which could lead to changes in the virulence modulation and positively affect infection treatment [10,14,15]. Regarding *spA* and *coa*, nisin-biogel at sub-MICs exhibited a trend to increase the expression of these genes, which can negatively affect infection treatment, as both SpA and Coa contribute for bacterial evasion from the host immune system [4,5]. Nisin-biogel at sub-MICs significantly decrease the expression of *atl* and *clfA*, possibly reducing *S. aureus* pathogenic potential. These results are in accordance with those by Zhao et al. [56], which stated that the exposure of *S. aureus* for 1 h to a nisin concentration equivalent to 1/2 MIC lead to a down-regulation of *atl* and *clfA*.

Concerning *icaA* and *icaD*, involved in biofilm formation, the nisin-biogel at subinhibitory concentrations slightly increased their expression, which can contribute for an increase in biofilm formation, hindering infection treatment. Moreover, the lower the concentration of nisin-biogel, a higher increase of *icaA* and *icaD* mRNA levels was observed, which reinforces the importance of defining the optimum dosage of antimicrobials to be applied to the treatment of bacterial infections.

The effects of subinhibitory concentrations on bacterial biofilm formation were also investigated. Biofilms play a major role in the pathogenesis of *S. aureus*, and are present in 60 to 100% of chronic wounds, including DFUs [57]. Several studies have shown that sub-MICs of some antimicrobials can affect bacterial biofilm formation in vitro, which may have clinical importance [58,59]. Angelopoulou et al. [60] observed that nisin at 1/2, 1/4 and 1/8 MIC significantly increased biofilm formation by *S. aureus*, with the concentration of 1/8 MIC also being the one that most influenced biofilm formation, as observed in this study. When exposed to nisin-biogel at 1/4 and 1/8 MIC values, the ability of *S. aureus* clinical isolates to form biofilm increased in a dose-dependent manner, while 1/2 MIC had no effect on biofilm formation. Moreover, the lower the concentration of nisin-biogel, the higher the increase in biofilm formation, which is mostly consistent with the effect on the expression of *icaA* and *icaD*. Moreover, as stated by other authors, the increase of biofilm formation may be due to the fact that sublethal concentrations of antimicrobials are cell stressors, which can enhance the production of biofilm matrix polymers [59,61]. However, Andre et al. [58] reported that subinhibitory concentrations of nisin promoted a reduction in biofilm formation, which suggests that different strains may present different responses regarding biofilm formation in the presence of sub-MICs of this antimicrobial peptide.

Alterations in the virulence determinants mRNA levels in the presence of subinhibitory levels of antimicrobials do not always result in changes in protein synthesis or functional activity [62,63]. According to our results, the increase in quantification of *spA* and *coa* mRNA in the presence of nisin-biogel at sub-MICs are not directly associated with an increase in SpA production or Coa activity, respectively. In fact, there was an unexpected significant increase of *coa* expression in the presence of subinhibitory levels of nisin-biogel. When accessing coagulase production, results were different, as most of the clinical isolates showed a similar coagulase activity in the presence of nisin-biogel at sub-MICs. Thus, our findings suggest that *coa* expression increase that occurs in the presence of nisin-biogel at sub-MICs may be associated with an increase in mRNA stabilization, as previously suggested by Blickwede et al. [64] regarding clindamycin influence on *coa* stability. On another end, SpA protein levels exhibited a trend to decrease in the presence of nisin-biogel at sub-MICs, which is partly not consistent with what happened to SpA mRNA levels. Actually, in the presence of nisin-biogel at sub-MICs levels, *S. aureus* DFI isolates increased SpA mRNA levels, which seems to suggest that the inhibition of virulence expression by nisin-biogel is primarily due to the blockage of protein translation at the ribosome, rather than the inhibition of virulence genes transcription.

4. Materials and Methods

4.1. Bacterial Strains

S. aureus isolates used in the present study were previously obtained from patients with clinically infected DFUs using biopsies (isolates with a B in their identification code), swabs (isolates with a Z in their identification code) or aspirates (isolates with an A in their identification code), according to the current clinical guidelines [65]. The isolates were virulence and antimicrobial resistance profiles were previsouly analyzed by Polymerase Chain Reaction (PCR), including their biofilm-forming ability, and their clonal profile was previously determined by Pulse Field Gel Electrophoresis (PFGE) and Multilocus Sequence Type (MLST) [66]. *S. aureus* isolates A 5.2, A 6.3, B 1.1, B 14.2, Z 1.1 and Z 5.2 were selected for this study due to their virulence traits, which includes the presence of genes encoding for staphylococcal protein A (*spA*), autolysin (*atl*), clumping factor A (*clfA*), coagulase (*coa*), intracellular adhesin A (*icaA*), intracellular adhesin D (*icaD*) and the accessory regulator gene I (*agrI*), and to their capacity to produce biofilms. Moreover, isolates B 14.2, B 1.1 and Z 1.1 are methicillin-resistant *S. aureus* (MRSA), whereas isolates A 5.2, A 6.3 and Z 5.2 are methicillin-susceptible *S. aureus* (MSSA) [38,66]. The MICs of nisin-biogel and clindamycin for the *S. aureus* clinical isolates were previously determined, being, in average, of 22.5 μg/mL and 0.033 μg/mL, respectively [27,67].

4.2. Antimicrobial Solutions

A stock solution of nisin (1000 μg/mL) was obtained by dissolving 1 g of nisin powder (2.5% purity, 1000 IU/mg, Sigma-Aldrich, St. Louis, MO, USA) in 25 mL of HCl (0.02 M) (Merck, Darmstadt, Germany). This solution was sterilized by filtration through a 0.22 μm cellulose acetate membrane filter (Millipore, Burlington, MA, USA) and stored at 4 °C [27].

Brain Heart Infusion (BHI) or Trypticase Soy Broth (TSB) with guar-gum gel at 0.75% (*w/v*) were prepared by dissolving 3.75 g of guar-gum (Sigma-Aldrich, St. Louis, MO, USA) and 18.5 g of BHI powder (VWR Chemicals, Leuven, Belgium) or 15 g of TSB powder (VWR Chemicals, Leuven, Belgium), respectively, in 500 mL of sterile distilled water, and heat sterilized by autoclave [27].

Clindamycin is an alternative antibiotic currently used in clinical practice associated with mild DFIs and was used in the present study at 1/2 MIC as a control for comparing the effects of nisin-biogel at sub-MICs on *S. aureus* DFI isolates virulence factors expression [28]. A stock solution of clindamycin was obtained by dissolving 6.6 mg of clindamycin powder (Sigma-Aldrich, St. Louis, MO, USA) in 10 mL of sterile water and filtered using a 0.22 μm cellulose acetate membrane filter. This stock solution was kept frozen at −80 °C and diluted with sterile water to the final concentration of 0.0165 μg/mL when required.

4.3. Effects of Nisin-Biogel Sub-MICs on S. aureus DFI Isolates Growth Rate

S. aureus DFI isolates were inoculated in a non-selective BHI agar medium (VWR Chemicals, Leuven, Belgium) at 37 °C for 24 h. After incubation, bacterial suspensions of 10^8 CFU/mL were prepared directly from plate cultures using a 0.5 McFarland standard in NaCl (Merck, Damstrants, Germany), and the bacterial suspensions were diluted in fresh BHI broth or in fresh BHI broth containing guar gum at 0.75% (*w/v*) to a final concentration of 10^7 CFU/mL. Afterwards, nisin was added to the fresh BHI broth with guar gum to obtain bacterial cultures with nisin-biogel at 1/2, 1/4 and 1/8 MIC values. Then, the wells of a 96-well flat-bottomed polystyrene microtiter plates (Thermo Scientific, Waltham, MA, USA) were inoculated with 200 μL of the negative controls (fresh BHI broth and fresh BHI broth with guar gum at 7.5 mg/mL) and with 200 μL of the different bacterial suspensions previously prepared, namely in fresh BHI broth, BHI broth with guar gum plus 1/2 MIC of the nisin-biogel ((nisin) = 11.25 μg/mL; (guar gum) = 7.5 mg/mL), BHI broth with guar gum plus 1/4 MIC of the nisin-biogel ((nisin) = 5.625 μg/mL; (guar gum) = 7.5 mg/mL) and BHI broth with guar gum plus 1/8 MIC of the nisin-biogel ((nisin) = 2.8175 μg/mL; (guar gum) = 7.5 mg/mL).

Each different growth condition was evaluated in triplicate wells on three independent assays. During the 24 h of incubation at 37 °C with shaking (150 rpm), optical density at 600 nm (OD_{600}) for each well were obtained automatically every hour, using the FLUOstar OPTIMA (BMG LABTECH, Ortenberg, Germany) microplate reader. For each isolate, OD_{600} was calculated by subtracting the average OD_{600} of the three blank wells (fresh BHI broth or fresh BHI broth with guar gum at 7.5 mg/mL) from the average of OD_{600} of the three replicates of the sample under evaluation. Final result corresponds to the average results of each independent assay replicates.

4.4. S. aureus DFI Isolates Gene Expression Kinetics

Isolates A 5.2 and Z 5.2 were inoculated in a non-selective BHI agar medium at 37 °C for 24 h. After incubation, bacterial suspensions of 10^8 CFU/mL were prepared as described. Bacterial suspensions were diluted in fresh BHI broth to a final concentration of 10^7 CFU/mL and grown at 37 °C with gyratory shaking (180 rpm). For quantification of *agrI*, *spA*, *coa*, *clfA* and *atl* expression, aliquots were collected after 2, 3, 4, 5 and 6 h of incubation for subsequent stabilization of total RNA with RNAprotect® Bacteria Reagent (Qiagen, Hilden, Germany), enzymatic lysis of bacteria with Buffer TE containing lysostaphin (Sigma-Aldrich, St. Louis, MO, USA), RNA extraction using the Qiagen RNeasy Mini Kit (Qiagen, Hilden, Germany), and cDNA synthesis with random primers using a Promega Go ScriptTM Reverse Transcription System (Promega, Madison, WI, USA), according to manufacturer's instructions. Gene expression was analyzed by RT-qPCR [68], using the specific primers shown in Table S6 (available as Supplementary Data). RT-qPCR was performed in a 7300 Real Time PCR System (Applied Biosystems, Waltham, MA, USA) using the following conditions: an initial uracil-N-glycosylase gene (UNG) activation at 50 °C for 2 min, followed by an initial DNA polymerase activation at 95 °C for 10 min, and 35 cycles consisting in melting at 95 °C for 15 s and annealing/extending at 60 °C for 1 min. A set of dissociation steps was also performed, using the following conditions: 95 °C for 15 s, 60 °C for 1 min, 95 °C for 15 s, and 60 °C for 15 s. For *icaA* and *icaD* genes, aliquots were collected after 8, 24, 32, 48 and 56 h of incubation for subsequent stabilization of total RNA, enzymatic lysis of bacteria, RNA extraction, cDNA synthesis and analyses by RT-qPCR. The relative standard curve method was used to quantify gene transcription, using gyrase B (*gyrB*) gene for normalization. An average ± standard deviation of the fold change obtained for the isolates A 5.2 and Z 5.2 was considered for the determination of gene expression kinetics and, consequently, the best growth time for further investigate the effects of nisin-biogel at sub-MICs on virulence-related genes expression.

4.5. Effects of Nisin-Biogel Sub-MICs on Gene Expression by S. aureus DFI Isolates

Isolates A 5.2, A 6.3, B 1.1, B 14.2, Z 1.1 and Z 5.2 were inoculated in a non-selective BHI agar medium at 37 °C for 24 h. After incubation, bacterial suspensions of 10^8 CFU/mL were prepared as described. For each isolate, 5 different bacterial suspensions were prepared: in BHI broth, BHI broth plus clindamycin at 1/2 MIC ((clindamycin) = 0.0165 µg/mL), BHI broth with guar-gum at 7.5 mg/mL plus 1/2 MIC of the nisin-biogel ((nisin) = 11.25 µg/mL), BHI broth with guar-gum at 7.5 mg/mL plus 1/4 MIC of the nisin-biogel ((nisin) = 5.625 µg/mL) and in BHI broth with guar-gum at 7.5 mg/mL plus 1/8 MIC of the nisin-biogel ((nisin) = 2.8175 µg/mL). All these suspensions were incubated at 37 °C for 4 h to *agrI*, *spa*, *atl*, *coa* and *clfa* genes expression studies, and for 48 h to *icaA* and *icaD* genes expression studies, with gyratory shaking (180 rpm).

After incubation, stabilization of total RNA, enzymatic lysis of bacteria, RNA extraction and cDNA synthesis were performed, and the resulting cDNA was used as a template for RT-qPCR using the specific primers shown in Table S1 (available as Supplementary Data), and the relative standard curve method was used to quantify transcription. Therefore, to determine the effects of nisin-biogel and clindamycin sub-MICs on virulence-related genes expression, the expression levels of the genes under investigation were expressed as fold change of the *spa/gyrB*, *agrI/gyrB*, *coa/gyrB*, *clfa/gyrB*, *icaA/gyrB*, *icaD/gyrB* and *atl/gyrB* ratios in the presence of antimicrobials (nisin-biogel or clindamycin) relative to the *spa/gyrB*, *agrI/gyrB*, *coa/gyrB*, *clfa/gyrB*, *icaA/gyrB*, *icaD/gyrB* and *atl/gyrB* ratios, respectively, of the growth control (no antimicrobial present). For each isolate and each incubation condition, two different and independent assays were performed.

4.6. Effect of Nisin-Biogel Sub-MICs on the Ability of S. aureus DFI Isolates to Form Biofilm

To test the influence of nisin-biogel sub-MICs on biofilm formation, a modified version of the protocol described by Santos et al. [27] was performed. Isolates A 5.2, A 6.3, B 1.1, B 14.2, Z 1.1 and Z 5.2 were inoculated in a non-selective BHI agar medium at 37 °C for 24 h. Then, three to five colonies were collected using a sterile loop, resuspended in 5 mL of TSB and incubated for 18 h at 37 °C. After incubation, the turbidity of bacterial suspension was adjusted to 0.5 McFarland standard (10^8 CFU/mL), and 1:100 dilutions were made in TSB with 0.25% glucose, TSB with 0.25% glucose plus 1/2 MIC of clindamycin ((clindamycin) = 0.0165 µg/mL), TSB with guar-gum at 7.5 mg/mL and 0.25% glucose plus 1/2 MIC of nisin-biogel ((nisin) = 11.25 µg/mL), TSB with guar-gum at 7.5 mg/mL and 0.25% glucose plus 1/4 MIC of nisin-biogel ((nisin) = 5.625 µg/mL) and TSB with guar-gum at 7.5 mg/mL and 0.25% glucose plus 1/8 MIC of nisin-biogel ((nisin) = 2.8175 µg/mL).

Bacterial suspensions were transferred to a sterile 96-well polystyrene plate (200 µL/well) and incubated at 37 °C for 48 h. After incubation, the content of each well was removed, and the wells were carefully washed three times with 180 µL of PBS, pH 7.0. Then, wells were filled with 200 µL of PBS, pH 7.0, and the microtiter plate was incubated in an ultrasound bath (Grant MXB14), at 50 Hz for 15 min, in order to disperse the biofilm-based bacteria from the microtiter plate surface. Finally, the OD of the suspension from each well was measured at 570 nm using the FLUOstar OPTIMA microplate reader. Results were calculated by subtracting the average OD_{570} of the three replicas of the negative controls (TSB broth or TSB broth with guar gum at 7.5 mg/mL) from the average of OD_{570} of the three wells of each sample. Final result corresponds to the average of the three independent assays.

4.7. Effect of Nisin-Biogel Sub-MICs on Coa Production by S. aureus DFI Isolates

Isolates A 5.2, A 6.3, B 1.1, B 14.2, Z 1.1 and Z 5.2 were inoculated in a non-selective BHI agar medium at 37 °C for 24 h. After, isolates were incubated in BHI broth, BHI broth plus clindamycin at 1/2 MIC ((clindamycin) = 0.0165 µg/mL), BHI broth with guar-gum at 7.5 mg/mL plus nisin-biogel at 1/2 MIC ((nisin) = 11.25 µg/mL), BHI broth with guar-gum at 7.5 mg/mL plus nisin-biogel at 1/4 MIC ((nisin) = 5.625 µg/mL) and BHI broth with guar-gum at 7.5 mg/mL plus nisin-biogel at 1/8 MIC ((nisin) = 2.8175 µg/mL) for 24 h or for 4 h at 37 °C. Coagulase test was performed by adding 0.1 mL of each culture to 0.3 mL of rabbit plasma previously rehydrated with sterile water. After gentle mixing, suspensions were incubated at 37 °C and examined every hour for 4 h, and after 24 h. Results were interpreted according to the scale proposed by Sperber & Tatini, 1975, where negative means no evidence of fibrin formation, positive 1+ means small unorganized clots, positive 2+ means small organized clot, positive 3+ means large organized clot, and positive 4+ means that the entire content of tube coagulates and is not displaced when tube is inverted [69]. As negative controls, 0.1 mL of BHI broth or 0.1 mL of BHI broth with guar-gum at 7.5 mg/mL were added to 0.3 mL of rabbit plasma and incubated without bacteria in the same conditions.

4.8. Effect of Nisin-Biogel Sub-MICs on SpA Release by S. aureus DFI Isolates

Isolates A 5.2, A 6.3, B 1.1, B 14.2, Z 1.1 and Z 5.2 were inoculated in a non-selective BHI agar medium at 37 °C for 24 h. After, isolates were incubated in BHI broth for 4 h, i.e., isolates were grown to the exponential phase at 37 °C. Then, the cultures were incubated in the presence of nisin-biogel at 1/2 MIC ((nisin) = 11.25 µg/mL), 1/4 MIC ((nisin) = 5.625 µg/mL) and 1/8 MIC ((nisin) = 2.8175 µg/mL), and clindamycin at 1/2 MIC ((clindamycin) = 0.0165 µg/mL) for 18 h at 37 °C with shaking. After incubation, bacterial suspensions were centrifuged at $1500 \times g$ rpm for 10 min at 4 °C, and supernatant were used to determine SpA level using the SpA ELISA Kit (Abcam, Cambridge, UK), as recommended by the manufacturer. Samples and standards were added to the 96-well plate, the assay was performed and absorbance values at 450 nm determined, as these values are directly proportional to the level of protein A in the sample. The results correspond to the ratios of the amount of SpA (pg/mL) in the bacterial supernatants incubated with clindamycin or nisin-biogel and the mean amount of SpA (pg/mL) in the bacterial supernatants incubated without antimicrobials and are expressed as percentages.

4.9. Statistical Analysis

Statistical analysis was carried out using Microsoft Excel 2016®. Quantitative variables are expressed as mean values ± standard deviation. Comparisons between treatments and control were performed using two-tailed Student's *t*-tests. A confidence interval of 95% was considered, and *p*-values < 0.05 indicate statistical significance.

5. Conclusions

S. aureus produces a wide variety of virulence factors, such as adherence and colonization molecules, exotoxins, and enzymes, and forms biofilms, which contribute to its ability to colonize host tissues and cause disease, making it difficult to control staphylococcal infections. The effect of antimicrobial agents on these virulence factors' expression has become a major focal point in the study of new antimicrobial alternatives. The present results demonstrated that nisin-biogel at subinhibitory levels affects the growth of *S. aureus* in a strain-dependent and dose-dependent manner, as well as the production of several virulence factors, including coagulase, protein A, and biofilm. The expression of some virulence-related genes, such as *agrI*, *atl* and *clfA*, were found to be repressed by nisin-biogel at sub-MICs, whereas the transcription levels of *spA*, *coa*, *icaA* and *icaD* were increased. Results highlight the importance of accessing the effects of nisin-biogel sub-MICs at different levels, providing an in vitro basis to understand what happens in vivo throughout the treatment of a DFI, and emphasizes how critical it is to establish the correct dosage of antimicrobials to be applied in clinical practice.

Supplementary Materials: The following are available online at https://www.mdpi.com/article/10.3390/antibiotics10121501/s1, Table S1: Primers used in RT-qPCR protocols using 7300 Real Time PCR System for accessing virulence gene expression; Table S2: Effects of nisin-biogel at 1/2 MIC, 1/4 MIC and 1/8 MIC and clindamycin at 1/2 MIC on *agrI*, *spA*, *atl*, *clfA*, *coa*, *icaA* and *icaD* mRNA expression for the isolates A 5.2, A 6.3, B 1.1, B 14.2, Z 1.1 and Z 5.2. Results are expressed as n-fold differences in 'gene under study/*gyrB*' ratio in the presence of the different conditions described above relative to 'gene under study/*gyrB*' ration in the growth control (no antimicrobial). Values are present as mean values ± SD (two repeated different experiments), except for the *clfa/gyrB* fold change for the isolate Z 5.2, since only one assay was performed. Asterisks indicate statistically significant differences between treatments and between treatments and control for each clinical isolate (* = *p* < 0.05; ** = *p* < 0.01; *** = *p* < 0.001, compared with the results of the corresponding control); Table S3: Effects of nisin-biogel at 1/2 MIC, 1/4 MIC and 1/8 MIC and clindamycin at 1/2 MIC on biofilm formation by the isolates A 5.2, A 6.3, B 1.1, B 14.2, Z 1.1 and Z 5.2. Values are present as mean values ± SD (three repeated different experiments). * = *p* < 0.05; ** = *p* < 0.01; *** = *p* < 0.001, compared with the results of the corresponding control; Table S4: Effects of nisin-biogel at 1/2 MIC, 1/4 MIC and 1/8 MIC and clindamycin at 1/2 MIC on coagulase production by the isolates A 5.2, A 6.3, B 1.1, B 14.2, Z 1.1 and Z 5.2 after 24 h of growth under the different conditions tested. Coagulation ability was measured every hour for 4 h of incubation, and after 24 h of incubation. -: no evidence of fibrin formation; 1+: small unorganized clots; 2+: small organized clots; 3+: large organized clots; 4+: entire content of tube coagulates and is not displaced when tube is inverted; Table S5: Effects of nisin-biogel at 1/2 MIC, 1/4 MIC and 1/8 MIC and clindamycin at 1/2 MIC on coagulase production by the isolates A 5.2, A 6.3, B 1.1, B 14.2, Z 1.1 and Z 5.2 after 4 h of growth under the different conditions tested. Coagulation ability was measured every hour for 4 h of incubation, and after 24 h of incubation. -: no evidence of fibrin formation; 1+: small unorganized clots; 2+: small organized clots; 3+: large organized clots; 4+: entire content of tube coagulates and is not displaced when tube is inverted; Table S6: Effects of nisin-biogel at 1/2 MIC, 1/4 MIC and 1/8 MIC and clindamycin at 1/2 MIC on protein A release by *S. aureus* DFI isolates A 5.2, A 6.3, B 1.1, B 14.2, Z 1.1 and Z 5.2 after 18 h of growth under the different conditions tested. The results are the racios of the amount of SpA (pg/mL) in the bacterial supernatants incubated with nisin-biogel or clindamycin to the amount of SpA (pg/mL) in the bacterial supernatants incubated without antimicrobials and are expressed as percentages.

Author Contributions: Conceptualization, C.J., R.S. and M.O.; Methodology, C.J., R.S. and M.O.; Validation, E.C., M.G., L.T. and M.O.; Investigation, C.J.; Data curation, C.J.; Writing—original draft preparation, C.J.; Writing—review and editing, R.S., E.C., M.G., L.T. and M.O.; Supervision, M.O.; Project administration, L.T. and M.O.; Funding acquisition, L.T. and M.O. All authors have read and agreed to the published version of the manuscript.

Funding: This research was funded by Foundation for Science and Technology, project PTDC/SAU-INF/28466/2017—Polyphasic validation of antimicrobial peptides as alternative treatment for diabetic foot infections). The authors would also like to recognize the Foundation of Science and Technology for the PhD fellowship SFRH/BD/131384/2017 (Eva Cunha) and to the University of Lisbon for the PhD fellowship C10571K (Miguel L. Grilo).

Institutional Review Board Statement: Not applicable.

Informed Consent Statement: Not applicable.

Data Availability Statement: The data presented in this study are available in the article or Supplementary Materials.

Conflicts of Interest: The authors declare no conflict of interest.

References

1. IDF Diabetes Atlas—9th Edition. Available online: https://diabetesatlas.org/ (accessed on 27 November 2021).
2. Armstrong, D.G.; Boulton, A.J.M.; Bus, S.A. Diabetic Foot Ulcers and Their Recurrence. *N. Engl. J. Med.* **2017**, *376*, 2367–2375. [CrossRef] [PubMed]
3. Noor, S.; Khan, R.U.; Ahmad, J. Understanding Diabetic Foot Infection and Its Management. *Diabetes Metab. Syndr. Clin. Res. Rev.* **2017**, *11*, 149–156. [CrossRef] [PubMed]
4. Wang, B.; Muir, T.W. Regulation of Virulence in *Staphylococcus aureus*: Molecular Mechanisms and Remaining Puzzles. *Cell Chem. Biol.* **2016**, *23*, 214–224. [CrossRef] [PubMed]
5. Periasamy, S.; Joo, H.S.; Duong, A.C.; Bach, T.H.L.; Tan, V.Y.; Chatterjee, S.S.; Cheung, G.Y.C.; Otto, M. How *Staphylococcus aureus* Biofilms Develop Their Characteristic Structure. *Proc. Natl. Acad. Sci. USA* **2012**, *109*, 1281–1286. [CrossRef] [PubMed]
6. Biswas, R.; Voggu, L.; Simon, U.K.; Hentschel, P.; Thumm, G.; Götz, F. Activity of the Major Staphylococcal Autolysin *Atl. FEMS Microbiol. Lett.* **2006**, *259*, 260–268. [CrossRef]
7. Porayath, C.; Suresh, M.K.; Biswas, R.; Nair, B.G.; Mishra, N.; Pal, S. Autolysin Mediated Adherence of *Staphylococcus aureus* with Fibronectin, Gelatin and Heparin. *Int. J. Biol. Macromol.* **2018**, *110*, 179–184. [CrossRef]
8. Singh, V.K. High Level Expression and Purification of *Atl*, the Major Autolytic Protein of *Staphylococcus aureus*. *Int. J. Microbiol.* **2014**, *2014*, 1–7.
9. Reffuveille, F.; Josse, J.; Vallé, Q.; Mongaret, C.; Gangloff, S.C. *Staphylococcus aureus* Biofilms and Their Impact on the Medical Field; InTech: London, UK, 2017; Volume 11, p. 187.
10. Dickschat, J.S. Quorum Sensing and Bacterial Biofilms. *Nat. Prod. Rep.* **2010**, *27*, 343–369. [CrossRef] [PubMed]
11. Maira-Litrán, T.; Kropec, A.; Abeygunawardana, C.; Joyce, J.; Mark, G.; Goldmann, D.A.; Pier, G.B. Immunochemical Properties of the Staphylococcal Poly-N-Acetylglucosamine Surface Polysaccharide. *Infect. Immun.* **2002**, *70*, 4433–4440. [CrossRef]
12. Otto, M. Staphylococcal Biofilms. In *Gram-Positive Pathogens*; ASM Press: Washington, DC, USA, 2019; pp. 699–711.
13. O'Gara, J.P. Ica and beyond: Biofilm Mechanisms and Regulation in *Staphylococcus epidermidis* and *Staphylococcus aureus*. *FEMS Microbiol. Lett.* **2007**, *270*, 179–188. [CrossRef]
14. Huntzinger, E.; Boisset, S.; Saveanu, C.; Benito, Y.; Geissmann, T.; Namane, A.; Lina, G.; Etienne, J.; Ehresmann, B.; Ehresmann, C.; et al. *Staphylococcus aureus* RNAIII and the Endoribonuclease III Coordinately Regulate Spa Gene Expression. *EMBO J.* **2005**, *24*, 824–835. [CrossRef]
15. Kong, K.F.; Vuong, C.; Otto, M. *Staphylococcus* Quorum Sensing in Biofilm Formation and Infection. *Int. J. Med. Microbiol.* **2006**, 133–139. [CrossRef]
16. Chastain, C.A.; Klopfenstein, N.; Serezani, C.H.; Aronoff, D.M. A Clinical Review of Diabetic Foot Infections. *Clin. Podiatr. Med. Surg.* **2019**, *36*, 381–395. [CrossRef] [PubMed]
17. Field, D.; O'Connor, R.; Cotter, P.D.; Ross, R.P.; Hill, C. In Vitro Activities of Nisin and Nisin Derivatives Alone and In Combination with Antibiotics against *Staphylococcus* Biofilms. *Front. Microbiol.* **2016**, *7*, 508. [CrossRef]

18. Grigoropoulou, P.; Eleftheriadou, I.; Jude, E.B.; Tentolouris, N. Diabetic Foot Infections: An Update in Diagnosis and Management. *Curr. Diabetes Rep.* **2017**, *17*, 3. [CrossRef] [PubMed]

19. Han, D.; Sherman, S.; Filocamo, S.; Steckl, A.J. Long-Term Antimicrobial Effect of Nisin Released from Electrospun Triaxial Fiber Membranes. *Acta Biomater.* **2017**, *53*, 242–249. [CrossRef]

20. Peterson, E.; Kaur, P. Antibiotic Resistance Mechanisms in Bacteria: Relationships between Resistance Determinants of Antibiotic Producers, Environmental Bacteria, and Clinical Pathogens. *Front. Microbiol.* **2018**, *9*, 2928. [CrossRef] [PubMed]

21. Perez, R.; Perez, M.T.; Elegado, F. Bacteriocins from Lactic Acid Bacteria: A Review of Biosynthesis, Mode of Action, Fermentative Production, Uses, and Prospects. *Int. J. Philipp. Sci. Technol.* **2015**, *8*, 61–67. [CrossRef]

22. Delves-Broughton, J.; Blackburn, P.; Evans, R.J.; Hugenholtz, J. Applications of the Bacteriocin, Nisin. *Antonie Van Leeuwenhoek* **1996**, *69*, 193–202. [CrossRef]

23. Punyauppa-Path, S.; Phumkhachorn, P. Nisin: Production and Mechanism of Antimicrobial Action. *Int. J. Curr. Res. Rev.* **2015**, *7*, 47–53.

24. Mirhosseini, H.; Amid, B.T. A Review Study on Chemical Composition and Molecular Structure of Newly Plant Gum Exudates and Seed Gums. *Food Res. Int.* **2012**, *46*, 387–398. [CrossRef]

25. Narsaiah, K.; Jha, S.N.; Wilson, R.A.; Mandge, H.M.; Manikantan, M.R. Optimizing Microencapsulation of Nisin with Sodium Alginate and Guar Gum. *J. Food Sci. Technol.* **2014**, *51*, 4054–4059. [CrossRef] [PubMed]

26. Prabaharan, M. Prospective of Guar Gum and Its Derivatives as Controlled Drug Delivery Systems. *Int. J. Biol. Macromol.* **2011**, *49*, 117–124. [CrossRef]

27. Santos, R.; Gomes, D.; Macedo, H.; Barros, D.; Tibério, C.; Veiga, A.S.; Tavares, L.; Castanho, M.; Oliveira, M. Guar Gum as a New Antimicrobial Peptide Delivery System against Diabetic Foot Ulcers *Staphylococcus aureus* Isolates. *J. Med. Microbiol.* **2016**, *65*, 1092–1099. [CrossRef] [PubMed]

28. Santos, R.; Ruza, D.; Cunha, E.; Tavares, L.; Oliveira, M. Diabetic Foot Infections: Application of a Nisin-Biogel to Complement the Activity of Conventional Antibiotics and Antiseptics against *Staphylococcus aureus* Biofilms. *PLoS ONE* **2019**, *14*, e0220000. [CrossRef]

29. Gomes, D.; Santos, R.; Soares, R.S.; Reis, S.; Carvalho, S.; Rego, P.; Peleteiro, M.C.; Tavares, L.; Oliveira, M. Pexiganan in Combination with Nisin to Control Polymicrobial Diabetic Foot Infections. *Antibiotics* **2020**, *9*, 128. [CrossRef]

30. Soares, R.S.; Santos, R.; Cunha, E.; Tavares, L.; Trindade, A.; Oliveira, M. Influence of Storage on the Antimicrobial and Cytotoxic Activities of a Nisin-biogel with Potential to be Applied to Diabetic Foot Infections Treatment. *Antibiotics* **2020**, *9*, 781. [CrossRef] [PubMed]

31. Andersson, D.I.; Hughes, D. Microbiological Effects of Sublethal Levels of Antibiotics. *Nat. Rev. Microbiol.* **2014**, *12*, 465–478. [CrossRef]

32. Li, J.; Xie, S.; Ahmed, S.; Wang, F.; Gu, Y.; Zhang, C.; Chai, X.; Wu, Y.; Cai, J.; Cheng, G. Antimicrobial Activity and Resistance: Influencing Factors. *Front. Pharmacol.* **2017**, *8*, 364. [CrossRef]

33. Otto, M.P.; Martin, E.; Badiou, C.; Lebrun, S.; Bes, M.; Vandenesch, F.; Etienne, J.; Lina, G.; Dumitrescu, O. Effects of Subinhibitory Concentrations of Antibiotics Factor Expression by Community-Acquired Methicillin-*Staphylococcus aureus*. *J. Antimicrob. Chemother.* **2013**, *68*, 1524–1532. [CrossRef] [PubMed]

34. Davies, J.; Spiegelman, G.B.; Yim, G. The World of Subinhibitory Antibiotic Concentrations. *Curr. Opin. Microbiol.* **2006**, *9*, 445–453. [CrossRef]

35. Herbert, S.; Barry, P.; Novick, R.P. Subinhibitory Clindamycin Differentially Inhibits Transcription of Exoprotein Genes in *Staphylococcus aureus*. *Infect. Immun.* **2001**, *69*, 2996–3003. [CrossRef] [PubMed]

36. Schilcher, K.; Andreoni, F.; Dengler Haunreiter, V.; Seidl, K.; Hasse, B.; Zinkernagel, A.S. Modulation of *Staphylococcus aureus* Biofilm Matrix by Subinhibitory Concentrations of Clindamycin. *Antimicrob. Agents Chemother.* **2016**, *60*, 5957–5967. [CrossRef] [PubMed]

37. IDF Europe Members—Portugal. Available online: https://idf.org/our-network/regions-members/europe/members/153-portugal.html (accessed on 27 November 2021).

38. Heravi, F.S.; Zakrzewski, M.; Vickery, K.; Armstrong, D.G.; Hu, H. Bacterial Diversity of Diabetic Foot Ulcers: Current Status and Future Prospectives. *J. Clin. Med.* **2019**, *8*, 1935. [CrossRef] [PubMed]

39. Reeks, B.Y.; Champlin, F.R.; Paulsen, D.B.; Scruggs, D.W.; Lawrence, M.L. Effects of Sub-Minimum Inhibitory Concentration Antibiotic Levels and Temperature on Growth Kinetics and Outer Membrane Protein Expression in *Mannheimia haemolytica* and *Heamophilus somnus*. *Can. J. Vet. Res.* **2005**, *69*, 1–10.

40. Zhanel, G.G.; Hoban, D.J.; Harding, G.K. Subinhibitory Antimicrobial Concentrations: A Review of In Vitro and In Vivo Data. *Can. J. Infect. Dis.* **1992**, *3*, 193–201. [CrossRef]

41. Pratten, J.; Foster, S.J.; Chan, P.F.; Wilson, M.; Nair, S.P. *Staphylococcus aureus* Accessory Regulators: Expression within Biofilms and Effect on Adhesion. *Microbes Infect.* **2001**, *3*, 633–637. [CrossRef]

42. Liu, D.; Li, Z.; Wang, G.; Li, T.; Zhang, L.; Tang, P. Virulence Analysis of *Staphylococcus aureus* in a Rabbit Model of Infected Full-Thickness Wound under Negative Pressure Wound Therapy. *Int. J. Gen. Mol. Microbiol.* **2018**, *111*, 161–170. [CrossRef]

43. Mottola, C.; Mendes, J.J.; Cristino, J.M.; Cavaco-Silva, P.; Tavares, L.; Oliveira, M. Polymicrobial Biofilms by Diabetic Foot Clinical Isolates. *Folia Microbiol.* **2016**, *61*, 35–43. [CrossRef] [PubMed]

44. Peng, H.L.; Novick, R.P.; Kreiswirth, B.; Kornblum, J.; Schlievert, P. Cloning, Characterization, and Sequencing of an Accessory Gene Regulator (*agr*) in *Staphylococcus aureus*. *J. Bacteriol.* **1988**, *170*, 4365–4372. [CrossRef]

45. Gómez, M.I.; O'Seaghdha, M.; Magargee, M.; Foster, T.J.; Prince, A.S. *Staphylococcus aureus* Protein A Activates TNFR1 Signaling through Conserved IgG Binding Domains. *J. Biol. Chem.* **2006**, *281*, 20190–20196. [CrossRef]

46. Yanagihara, K.; Tashiro, M.; Fukuda, Y.; Ohno, H.; Higashiyama, Y.; Miyazaki, Y.; Hirakata, Y.; Tomono, K.; Mizuta, Y.; Tsukamoto, K.; et al. Effects of Short Interfering RNA against Methicillin-Resistant *Staphylococcus aureus* Coagulase In Vitro and In Vivo. *J. Antimicrob. Chemother.* **2006**, *57*, 122–126. [CrossRef]

47. Vandenesch, F.; Kornblum, J.; Novick, R.P. A Temporal Signal, Independent of Agr, Is Required for Hla but Not Spa Transcription in *Staphylococcus aureus*. *J. Bacteriol.* **1991**, *173*, 6313–6320. [CrossRef]

48. Lebeau, C.; Vandenesch, F.; Greenland, T.; Novick, R.P.; Etienne, J. Coagulase Expression in *Staphylococcus aureus* Is Positively and Negatively Modulated by an *agr*-Dependent Mechanism. *J. Bacteriol.* **1994**, *176*, 5534–5536. [CrossRef] [PubMed]

49. Farnsworth, C.W.; Schott, E.M.; Jensen, S.E.; Zukoski, J.; Benvie, A.M.; Refaai, M.A.; Kates, S.L.; Schwarz, E.M.; Zuscik, M.J.; Gill, S.R.; et al. Adaptive Upregulation of Clumping Factor A (*clfA*) by *Staphylococcus aureus* in the Obese, Type 2 Diabetic Host Mediates Increased Virulence. *Infect. Immun.* **2017**, *85*, e01005-16. [CrossRef] [PubMed]

50. Josefsson, E.; Kubica, M.; Mydel, P.; Potempa, J.; Tarkowski, A. In Vivo Sortase A and Clumping Factor A MRNA Expression during *Staphylococcus aureus* Infection. *Microb. Pathog.* **2008**, *44*, 103–110. [CrossRef] [PubMed]

51. Pasztor, L.; Ziebandt, A.K.; Nega, M.; Schlag, M.; Haase, S.; Franz-Wachtel, M.; Madlung, J.; Nordheim, A.; Heinrichs, D.E.; Götz, F. Staphylococcal Major Autolysin (Atl) Is Involved in Excretion of Cytoplasmic Proteins. *J. Biol. Chem.* **2010**, *285*, 36794–36803. [CrossRef]

52. Oshida, T.; Takano, M.; Sugai, M.; Suginaka, H.; Matsushita, T. Expression Analysis of the Autolysin Gene (*atl*) of *Staphylococcus aureus*. *Microbiol. Immunol.* **1998**, *42*, 655–659. [CrossRef]

53. Atshan, S.S.; Shamsudin, M.N.; Karunanidhi, A.; van Belkum, A.; Lung, L.T.T.; Sekawi, Z.; Nathan, J.J.; Ling, K.H.; Seng, J.S.C.; Ali, A.M.; et al. Quantitative PCR Analysis of Genes Expressed during Biofilm Development of Methicillin Resistant *Staphylococcus aureus* (MRSA). *Infect. Genet. Evol.* **2013**, *18*, 106–112. [CrossRef] [PubMed]

54. Patel, J.D.; Colton, E.; Ebert, M.; Anderson, J.M. Gene Expression during *S. epidermidis* Biofilm Formation on Biomaterials. *J. Biomed. Mater. Res.* **2012**, *100*, 2863–2869. [CrossRef]

55. Smieja, M. Current Indications for the Use of Clindamycin: A Critical Review. *Can. J. Infect. Dis.* **1998**, *9*, 22–28. [CrossRef]

56. Zhao, X.; Meng, R.; Shi, C.; Liu, Z.; Huang, Y.; Zhao, Z.; Guo, N.; Yu, L. Analysis of the Gene Expression Profile of *Staphylococcus aureus* Treated with Nisin. *Food Control* **2016**, *59*, 499–506. [CrossRef]

57. Malone, M.; Bjarnsholt, T.; McBain, A.J.; James, G.A.; Stoodley, P.; Leaper, D.; Tachi, M.; Schultz, G.; Swanson, T.; Wolcott, R.D. The Prevalence of Biofilms in Chronic Wounds: A Systematic Review and Meta-Analysis of Published Data. *J. Wound Care* **2017**, *26*, 20–25. [CrossRef] [PubMed]

58. Andre, C.; de Jesus Pimentel-Filho, N.; de Almeida Costa, P.M.; Vanetti, M.C.D. Changes in the Composition and Architecture of Staphylococcal Biofilm by Nisin. *Braz. J. Microbiol.* **2019**, *50*, 1083–1090. [CrossRef]

59. Kaplan, J.B. Antibiotic-Induced Biofilm Formation. *Int. J. Artif. Organs* **2011**, *34*, 737–751. [CrossRef] [PubMed]

60. Angelopoulou, A.; Field, D.; Pérez-Ibarreche, M.; Warda, A.K.; Hill, C.; Ross, R.P. Vancomycin and Nisin A Are Effective against Biofilms of Multi-Drug Resistant *Staphylococcus aureus* Isolates from Human Milk. *PLoS ONE* **2020**, *15*, e0233284. [CrossRef]

61. Majidpour, A.; Fathizadeh, S.; Afshar, M.; Rahbar, M.; Boustanshenas, M.; Heidarzadeh, M.; Arbabi, L.; Soleymanzadeh Moghadam, S. Dose-Dependent Effects of Common Antibiotics Used to Treat *Staphylococcus aureus* on Biofilm Formation. *Iran. J. Pathol.* **2017**, *12*, 362–370. [CrossRef]

62. Hodille, E.; Rose, W.; Diep, B.A.; Goutelle, S.; Lina, G.; Dumitrescu, O. The Role of Antibiotics in Modulating Virulence in *Staphylococcus aureus*. *Clin. Microbiol. Rev.* **2017**, *30*, 887–917. [CrossRef]

63. McAdow, M.; Missiakas, D.M.; Schneewind, O. *Staphylococcus aureus* Secretes Coagulase and von Willebrand Factor Binding Protein to Modify the Coagulation Casca de and Establish Host Infections. *J. Innate Immun.* **2012**, *4*, 141–148. [CrossRef]

64. Blickwede, M.; Wolz, C.; Valentin-Weigand, P.; Schwarz, S. Influence of Clindamycin on the Stability of Coa and FnbB Transcripts and Adherence Properties of *Staphylococcus aureus* Newman. *FEMS Microbiol. Lett.* **2005**, *252*, 73–78. [CrossRef]

65. Mendes, J.J.; Marques-Costa, A.; Vilela, C.; Neves, J.; Candeias, N.; Cavaco-Silva, P.; Melo-Cristino, J. Clinical and Bacteriological Survey of Diabetic Foot Infections in Lisbon. *Diabetes Res. Clin. Pract.* **2012**, *95*, 153–161. [CrossRef] [PubMed]

66. Mottola, C.; Semedo-Lemsaddek, T.; Mendes, J.J.; Melo-Cristino, J.; Tavares, L.; Cavaco-Silva, P.; Oliveira, M. Molecular Typing, Virulence Traits and Antimicrobial Resistance of Diabetic Foot Staphylococci. *J. Biomed. Sci.* **2016**, *23*, 33. [CrossRef]

67. Mottola, C.; Matias, C.S.; Mendes, J.J.; Melo-Cristino, J.; Tavares, L.; Cavaco-Silva, P.; Oliveira, M. Susceptibility Patterns of *Staphylococcus aureus* Biofilms in Diabetic Foot Infections. *BMC Microbiol.* **2016**, *16*, 119. [CrossRef]

68. Kearns, A.M.; Seiders, R.R.; Wheeler, J.; Freeman, R.; Steward, M. Rapid Detection of Methicillin-Resistant Staphylococci by Multiplex PCR. *J. Hosp. Infect.* **1999**, *43*, 33–37. [CrossRef] [PubMed]

69. Sperber, W.H.; Tatini, S.R. Interpretation of the Tube Coagulase Test for Identification of *Staphylococcus aureus*. *Appl. Microbiol.* **1975**, *29*, 502–505. [CrossRef] [PubMed]

Article

Characterization of bla_{KPC-2}-Carrying Plasmid pR31-KPC from a *Pseudomonas aeruginosa* Strain Isolated in China

Min Yuan [1,†], Hongxia Guan [2,†], Dan Sha [2], Wenting Cao [2], Xiaofeng Song [1], Jie Che [1], Biao Kan [1] and Juan Li [1,*]

[1] State Key Laboratory for Infectious Diseases Prevention and Control, Collaborative Innovation Center for Diagnosis and Treatment of Infectious Disease, National Institute for Communicable Disease Control and Prevention, Chinese Center for Disease Control and Prevention, Beijing 102206, China; yuanmin@icdc.cn (M.Y.); songxiaofeng@icdc.cn (X.S.); chejie@icdc.cn (J.C.); kanbiao@icdc.cn (B.K.)

[2] Wuxi Center for Disease Control and Prevention, Wuxi 214023, China; ghx-331@163.com (H.G.); shadan20051001@163.com (D.S.); caowent1978@126.com (W.C.)

* Correspondence: lijuan@icdc.cn; Tel.: +86-10-58900766

† These authors contributed equally to this work.

Abstract: This work aimed to characterize a 29-kb bla_{KPC-2}-carrying plasmid, pR31-KPC, from a multidrug resistant strain of *Pseudomonas aeruginosa* isolated from the sputum of an elderly patient with multiple chronic conditions in China. The backbone of pR31-KPC is closely related to four other bla_{KPC-2}-carrying plasmids, YLH6_p3, p1011-KPC2, p14057A, and pP23-KPC, none of which have been assigned to any of the known incompatibility groups. Two accessory modules, the IS26-bla_{KPC-2}-IS26 unit and IS26-ΔTn6376-IS26 region, separated by a 5.9-kb backbone region, were identified in pR31-KPC, which was also shown to carry the unique resistance marker bla_{KPC-2}. A comparative study of the above five plasmids showed that p1011-KPC2 may be the most complete plasmid of this group to be reported, while pR31-KPC is the smallest plasmid having lost most of its conjugative region. Regions between the iterons and *orf207* in the backbone may be hot spots for the acquisition of exogenous resistance entities. The accessory regions of these plasmids have all undergone several biological events when compared with Tn6296. The further transfer of bla_{KPC-2} in these plasmids may be initiated by either the Tn3 family or IS26-associated transposition or homologous recombination. The data presented here will contribute to a deeper understanding of bla_{KPC-2} carrying plasmids in *Pseudomonas*.

Keywords: *Pseudomonas aeruginosa*; carbapenem resistance; KPC-2; plasmid

Citation: Yuan, M.; Guan, H.; Sha, D.; Cao, W.; Song, X.; Che, J.; Kan, B.; Li, J. Characterization of bla_{KPC-2}-Carrying Plasmid pR31-KPC from a *Pseudomonas aeruginosa* Strain Isolated in China. *Antibiotics* **2021**, *10*, 1234. https://doi.org/10.3390/antibiotics10101234

Academic Editors: Manuela Oliveira, Elisabete Silva and Jonathan Frye

Received: 11 August 2021
Accepted: 9 October 2021
Published: 11 October 2021

1. Introduction

Pseudomonas aeruginosa is ubiquitous and is a well-known opportunistic pathogen in hospitalized, immunocompromised patients. *P. aeruginosa* is capable of infecting all types of systems and tissues, causing various diseases, such as bacteremia, septicemia, septicopyemia, pneumonia, bronchitis, diarrhea, keratitis, and skin and wound infections [1,2]. Multidrug resistant (MDR) *P. aeruginosa* has been categorized as a serious health threat by the Centers for Disease Control and Prevention since 2019 (https://www.cdc.gov/drugresistance/biggest_threats.html), and the development of resistance to carbapenem will further exacerbate the situation [3].

Resistance to carbapenemase in *P. aeruginosa* is generally due to a combination of mechanisms, including porin (OprD) deficiency, overexpression of efflux pumps, intrinsic chromosomally encoded AmpC-lactamase, and/or carbapenemase production [4]. Carbapanemase-encoding genes are highly transferable as they always reside on mobile genetic elements (MGEs) such as plasmids, gene cassettes of integrons, transposons, and genomic islands [5]. To date, the MGE-related carbapenemase families that have been reported in *P. aeruginosa* include class A (KPC-2 [6] and GES [7]), class B (IMP [8], VIM [9], SPM [10], NDM [11], AIM [12], SIM [13], FIM [14], HMB [15], and CAM [16]), and class D

(OXA-40-like [17], OXA-48-like [18], and OXA-198-like [19]). *P. aeruginosa* is the predominant host for class B enzymes, while class A and D enzymes are less commonly reported in this species [20].

The KPC-2-producing *P. aeruginosa* isolates were first reported in 2006 in a strain from Colombia [6] and have recently been reported in strains from North America [21], South America (Trinidad and Tobago [22], South Florida [23], and Puerto Rico [24]), China [25], Brazil [26], and Germany [27]. The bla_{KPC-2} gene can be either plasmid-borne or on the chromosome of the host. Using the keywords "pseudomonas", "KPC", "plasmid", and "complete" to search the GenBank database (last accessed on 14 May 2021), a total of six fully sequenced and published bla_{KPC-2}-carrying plasmids from *P. aeruginosa* were obtained: Two IncP-6 plasmids, pCOL-1 (accession number KC609323; from Colombia) [28] and p10265-KPC (accession number KU578314; from China) [29]; one IncU plasmid, pPA-2 (accession number KC609322; from Colombia) [28]; plasmid pBH6 (accession number CP029714; from Brazil) [30]; and p1011-KPC2 and p14057A (accession numbers MH734334, KY296095, and CP065418; all from China) [31,32], which all belong to the same unknown incompatibility group.

In this work, we characterized the bla_{KPC-2}-carrying *P. aeruginosa* plasmid pR31-KPC and compared its sequence with those of the sequenced plasmids in the database to gain a deeper understanding of the evolutionary history of this group of plasmids. Tn*6774* was the novel transposon designated in this study.

2. Results and Discussion

2.1. Case Report

On 31 August 2015, a 79-year-old man was admitted to a local hospital in Yixing City with a cough and expectoration that had persisted for half a year, which showed progressive aggravation for 20 h. The patient was subsequently diagnosed with pneumonia, hypertension, sequela of cerebral apoplexy, and post tracheotomy. The patient received a series of symptomatic treatments, such as aspiration of sputum to relieve the cough, oxygen inhalation to improve circulation, and intravenous administration of meropenem and tigecycline to reduce inflammation. *P. aeruginosa* R31 was isolated from a sputum specimen during hospitalization. On the eighteenth day of hospitalization, his symptoms worsened and he displayed shortness of breath, decreased blood pressure, decreased urine output, and symptoms of multiple organ dysfunction.

2.2. General Features of P. aeruginosa R31

Strain R31 was highly resistant to all of the β-lactams tested, including penicillins (piperacillin, piperacillin/tazobactam), cephalosporins (ceftazidime, cefepime), carbapenems (imipenem, meropenem), and monobactam (aztreonam). Moreover, it showed resistance to some fluoroquinolones (ciprofloxacin, levofloxacin), but was still susceptible to aminoglycosides (gentamicin, amikacin, and tobramycin) and colistin (Table 1).

Table 1. Minimum inhibitory concentration (MIC) values of *P. aeruginosa* R31 determined by the microdilution method.

	MIC Values		MIC Breakpoints µg/mL		
	µg/mL	R or S	S	I	R
Piperacillin	>1024	R	$\leq$16	32–64	$\geq$128
Piperacillin/tazobactam	>1024/4	R	$\leq$16/4	32/4–64/4	$\geq$128/4
Ceftazidime	128	R	$\leq$8	16	$\geq$32
Cefepime	>512	R	$\leq$8	16	$\geq$32
Imipenem	>128	R	$\leq$2	4	$\geq$8
Meropenem	>128	R	$\leq$2	4	$\geq$8
Aztreonam	>512	R	$\leq$8	16	$\geq$32

Table 1. *Cont.*

	MIC Values		MIC Breakpoints µg/mL		
	µg/mL	R or S	S	I	R
Gentamicin	4	S	≤ 4	8	≥ 16
Amikacin	<8	S	≤ 16	32	≥ 64
Tobramycin	<2	S	≤ 4	8	≥ 16
Ciprofloxacin	8	R	≤ 0.5	1	≥ 2
Levofloxacin	32	R	≤ 1	2	≥ 4
Colistin	1	S	≤ 2	-	≥ 4

The R31 strain returned a positive result in the Carba NP test, and out of all the carbapenemase genes tested, only bla_{KPC-2} was detected. Repeated conjugation experiments failed to transfer the bla_{KPC-2} marker from R31 to *P. aeruginosa* PAO1 (induced rifampin resistance) or Escherichia coli EC600 (rifampin resistance).

2.3. Overview of pR31-KPC

The R31 isolate harbors only one extrachromosomal closed circular DNA sequence, designated as pR31-KPC, which was determined to be 29,402 bp in length and contain a mean G+C content of 57.5% and 44 predicted open reading frames (ORFs) (Figure S1). The backbone of pR31-KPC has a modular structure, with the insertion of two accessory modules: The IS26-bla_{KPC-2}-IS26 unit and IS26-ΔTn6376-IS26 region. The accessory modules were defined as acquired DNA regions associated with and bordered by mobile elements.

2.4. The Backbone of pR31-KPC

The backbone of pR31-KPC is 16.9-kb in length and contains the following elements: The RepA and its iterons (repeat region for the RepA binding site), which are responsible for plasmid replication initiation. The iterons are 137 bp in size, within which 12-bp sites are located, relatively conserved, and repeated six times; parA for plasmid partition; higBA, which encodes the toxin-antitoxin system for post-segregational killing; and a traMLI conjugation system remnant. The identified RepA protein of pR31-KPC showed a 100% amino acid similarity to the homologs in the four other bla_{KPC-2}-carrying Pseudomonas plasmids of the same incompatibility group, which are available in public sequence databases, namely p1011-KPC2, p14057A, YLH6_P3 (accession number MK882885), and pP23-KPC (accession number CP065418).

The backbone of pR31-KPC showed 83–100% coverage and 100% identity to the above-mentioned plasmids (Supplementary Table S1). A linear comparison of the backbones of these five plasmids revealed the following: (1) The regions between the iterons and orf207 are hot spots for the acquisition of resistance genes, and all of the bla_{KPC-2} genes reside in these regions; (2) p1011-KPC2 is the most complete plasmid of this incompatibility group, with a complete conjugative region and a relatively intact maintenance region, while pR31-KPC is the smallest plasmid of this group (Figure 1). Although most of its conjugative region is missing and is unable to conjugate experimentally, pR31-KPC and thus, bla_{KPC-2} can remain in its host.

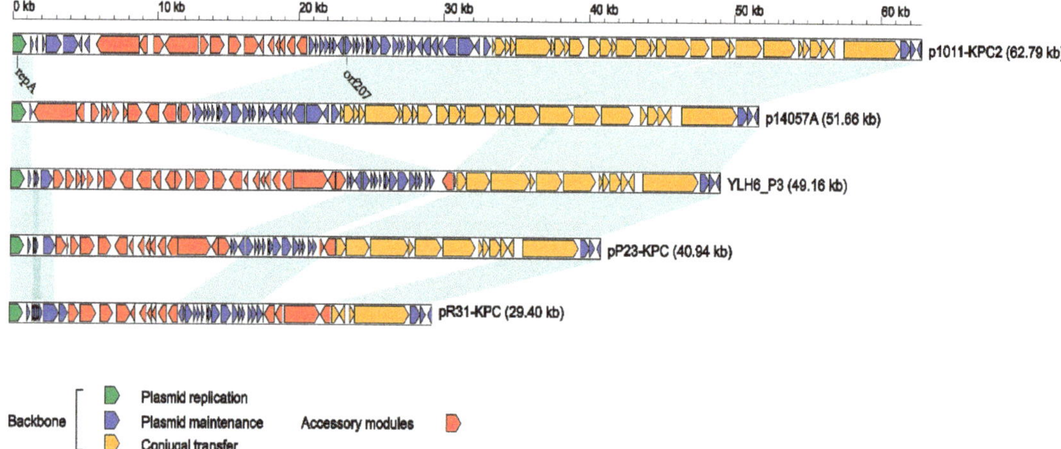

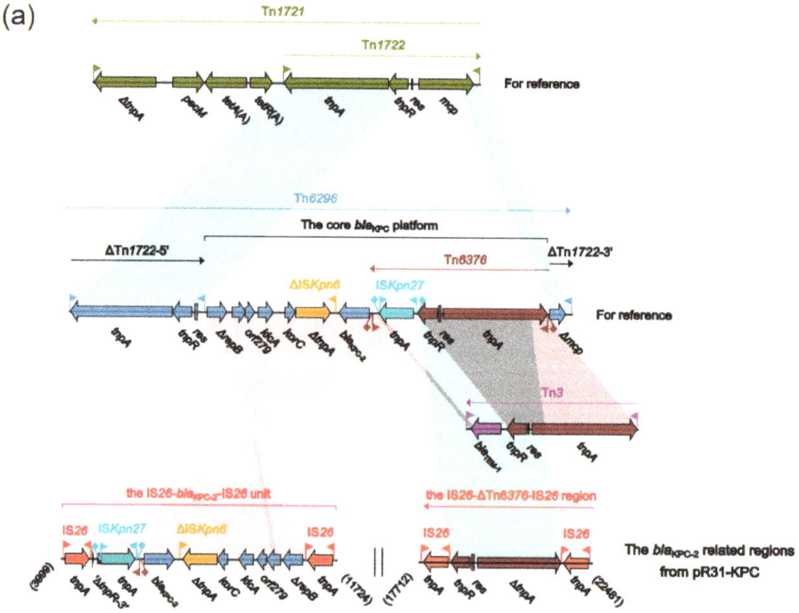

Figure 1. Linear comparison of plasmid genome sequences. Genes are denoted by arrows. The plasmid backbone replication, maintenance, and conjugation regions are colored in green, dark blue, and orange, respectively. The accessory module regions are colored in red. Shading denotes homology (nucleotide identity $\geq$ 90%) of the plasmid backbone regions, but not the accessory modules. *RepA* and *orf207* represent the names of the labeled genes, respectively. The GenBank accession numbers of p1011-KPC2, p14057A, YLH6_P3, pP23-KPC, and pR31-KPC are MH734334, KY296095, MK882885, CP065418, and CP061851, respectively.

2.5. The Accessory Regions of pR31-KPC

The accessory regions of pR31-KPC comprise two IS26-based regions, the IS26-*bla*$_{KPC-2}$-IS26 unit and IS26-ΔTn6376-IS26 region, separated by a backbone region of 5988 bp (Figure 2a).

Figure 2. *Cont.*

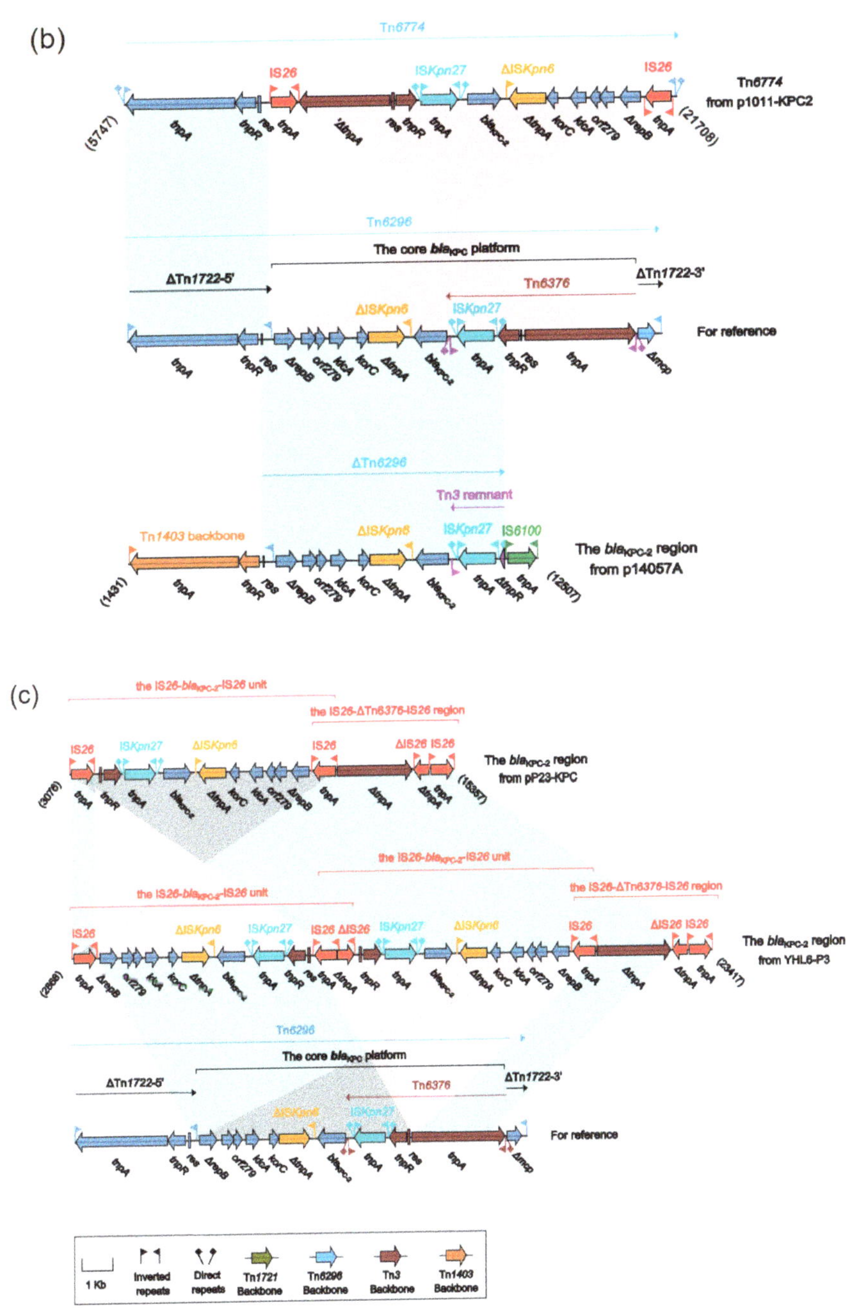

Figure 2. (a) The accessory regions pR31-KPC, and comparison with Tn*6296* and Tn*1721*; (b) The accessory regions p1011-KPC2 and p14057A, and comparison with Tn*6296*; (c) The accessory regions of pP23-KPC and YLH6_P3, and comparison with Tn*6296*. Genes are denoted by arrows. Genes, mobile elements, and other features are colored based on function classification. Shading denotes regions of homology (>95% nucleotide identity). Numbers in brackets indicate the nucleotide positions within the corresponding plasmids. The GenBank accession numbers of Tn*1721*, Tn*3*, Tn*6296*, and Tn*1403* are X61367, HM749966, FJ628167, and AF313472, respectively.

Tn6296 is widely considered to be one of the most important vehicles for bla_{KPC-2} gene transferring. Tn6296 was originally identified in MDR plasmid pKP048 from *Klebsiella pneumoniae*. In addition, it was generated from the insertion of the core bla_{KPC-2} genetic platform (Tn6376–bla_{KPC-2}–ΔISKpn6–korC–orf6–klcA–ΔrepB) into Tn1722, resulting in truncation of mcp (Figure 2a).

In pR31-KPC, the aforementioned core bla_{KPC-2} genetic platform is intact, but has been split into two parts, each of which is bordered by two IS26 elements (either in the same or opposite directions), generating the IS26-bla_{KPC-2}-IS26 unit and IS26-ΔTn6376-IS26 region, which have the potential to move (Figure 2a). Both of the regions lack the typical 5 bp target site duplications, suggesting that the acquisition of these entities may have occurred via the IS26-mediated homologous recombination.

In p1011-KPC2, two copies of IS26 were found at the boundaries of the core bla_{KPC-2} genetic platform in opposite directions, translocating the core platform and truncating $tnpA_{Tn6376}$ into a 2455 bp fragment. Regarding the integrity of Tn6296, the left/right inverted repeats and direct repeats were not impaired, generating the novel transposon Tn6774 (Figure 2b). The further spread of bla_{KPC-2} may occur by either the Tn6774 transposition via a TnpA/$TnpR_{Tn6774}$-mediated 'cut and paste' process or IS26-mediated transposition.

In p14057A, Tn6296 was truncated by the Tn1403 core tni module and IS6100 at either ends, generating the ΔTn1403-ΔTn6296-IS6100 region (Figure 2b). This entity may have been generated by a recombination of Tn6296 and a Tn1403-like transposon at the res site. Tn1403, initially found in Pseudomonas, is an important resistance gene dissemination vehicle, with the derivatives Tn6060, Tn6061, Tn6217, Tn6249, and Tn6286 having been reported in [33–37]. Belonging to the Tn21 subfamily of the Tn3 family, the Tn1403 and Tn1403-like transposons are able to transfer their passengers by the one-end transposition [38].

In YLH6_P3 and pP23-KPC, six copies of IS26 (four intact and two truncated) and four copies of IS26 (three intact and one truncated) were found in the bla_{KPC-2} region, respectively, forming mosaic structures, with adjacent IS26 regions overlapping each other. In YLH6_P3, two copies of bla_{KPC-2} were found. This structure was likely generated by the duplication of IS26-bla_{KPC-2}-IS26 found in pP23-KPC or vice versa. Linkage to IS26 indicates the potential for further dissemination of bla_{KPC-2} (Figure 2c).

2.6. Genomic Characterization of P. aeruginosa Genomes

A total of 209 genomes were downloaded (including that of R31) from the GenBank database. The resistance genes carried by each genome are listed in Table S2. All of the genomes, except for seven, have the chromosome-origin bla_{PAO-1} gene. The seven genomes without the gene were excluded for further whole genome phylogeny studies, due to the probable misidentification at the genus or species level. Phylogeny studies of the remaining 202 genomes revealed the following: (1) These clones can be divided into three clusters (clusters I, II, and III), and cluster III can be further divided into IIIa and IIIb. R31 belongs to cluster IIIa. (2) The carriage rates of carbapenemase genes for clusters I, II, IIIa, and IIIb were 33.3% (3/9), 36.45% (35/96), 23% (8/26), and 2.8% (2/71), respectively. Compared with clusters I, II, and IIIa, cluster IIIb clones have the lowest carriage rate of carbapenemase genes. (3) Worldwide, the sequence types (STs) of sequenced *P. aeruginosa* genomes are highly diverse.

The 202 genomes included a total of 69 known STs and 36 unknown STs. Forty-eight isolates with 23 different kinds of carbapenemase genotypes included 17 known STs and nine unknowns STs. A specific relationship between the STs and carbapenemase genotypes was not obvious (Figure S2).

3. Materials and Methods

3.1. Ethics Statement

The specimens were acquired with consent from the patient. The use of human specimens and all of the related experimental protocols was reviewed and approved by the ethics committee of the National Institute for Communicable Disease Control and

Prevention (ICDC), Beijing, China, in accordance with the medical research regulations of the Ministry of Health, China. Research involving biohazardous materials and all of the related procedures were approved by the Biosafety Committee of the ICDC. This study was conducted in China.

3.2. Identification of Bacterial Strains

Bacterial species were identified with the VITEK-2 Compact system using the GNI card (bioMerieux, France) and further confirmed by sequencing of the 16S rDNA amplicon, which is generated by primer pairs 27f (5′-AGAGTTTGATCCTGGCTCAG-3′) and 1492r (5′-ACGGCTACCTTGTTACGACTT-3′).

3.3. Determination of Minimum Inhibitory Concentration (MIC)

Antimicrobial susceptibility testing was performed by a broth microdilution method with customized microtiter plates containing vacuum dried antibiotics (BD Bioscience, San Jose, CA, USA). The MIC values were interpreted according to the Clinical and Laboratory Standards Institute (CLSI) guidelines 2019.

3.4. Detection of Carbapenemase Activity and Screening of Responsible Genes

The Carba NP test recommended by CLSI was performed for the detection of the carbapenemase production. The major plasmid-borne carbapenemase genes were amplified by the polymerase chain reaction (PCR) and sequenced on an ABI 3730 DNA Analyzer (Applied Biosystems, Foster City, CA, USA).

3.5. Conjugation Experiments

Conjugation experiments were carried out by a filter mating method with *P. aeruginosa* R31 as the donor strain and *P. aeruginosa* PAO1 (induced rifampin resistance) or *E. coli* EC600 (rifampin resistance) serving as the recipient strain. Briefly, the donor and recipient strains were grown in 3 mL of brain heart infusion (BHI) broth overnight at 37 °C. For each conjugation, 50 μL of donor strain culture was mixed with 500 μL of recipient strain culture (v:v = 1:10) and 4.5 mL of fresh BHI broth. In addition, 100 μL of the mixture was applied onto a cellulose filter membrane (pore size, 0.22 μm) already placed on a BHI agar plate. After incubation at 37 °C for 16–18 h, the filter membrane was taken out and vortexed in 1 mL of BHI broth. The vortex mixtures were plated on BHI agar plates containing 100 mg/L ceftazidime and 50 mg/L rifampin for the selection of the *P. aeruginosa* PAO1 transconjugants or on BHI agar plates containing 100 mg/L ceftazidime and 100 mg/L sodium azide for the selection of the *E. coli* EC600 transconjugants.

3.6. Determination of R31 Genome Sequences

Genomic DNA was isolated from the R31 isolate using a Wizard Genomic DNA Purification Kit (Promega, Madison, WI, USA), and sequenced using a whole-genome shotgun strategy on an Ion Torrent Personal Genome Machine system (Life Technologies). A Pacific Biosciences RSII DNA sequencing system (Menlo Park, CA, USA) was used as the platform to sequence the complete genome. The contigs were assembled using Hgap 2.0.

3.7. Sequence Annotation and Comparison

Sequence annotation and prediction of open reading frames (ORF) and pseudogenes of plasmid pR31-KPC were performed using RAST 2.0 [39] and further annotated by BLASTP/BLASTN [40] searches against the UniProtKB/Swiss-Prot [41] and RefSeq [42] databases. Annotation of resistance genes, mobile elements, and other features was conducted using online databases, such as CARD [43], ResFinder [44], ISfinder [45], ISsaga [46], INTEGRALL [47], and the Tn Number Registry [48]. Multiple and pairwise sequence comparisons were performed using MUSCLE 3.8.31 [49] and BLASTN, respectively. Gene organization diagrams were drawn in Inkscape 0.48.1 (http://inkscape.org, accessed on 1 January 2019)

3.8. Whole Genome Phylogeny and Genetic Background Analysis

The available whole genome sequences of *P. aeruginosa* from the National Center for Biotechnology Information (NCBI) were downloaded (last accessed on 24 October 2019). Resistance genes of the Pseudomonas genomes were analyzed by ResFinder. The sequence types (ST) of these *P. aeruginosa* genomes were obtained using the webtool PubMLST (https://pubmlst.org/organisms/pseudomonas-aeruginosa). Analyses of the downloaded genomes (including R31) were further processed using kSNP3 v3.1 [50]. Core single-nucleotide polymorphism (SNP) matrices were generated, and maximum-likelihood phylogenies were constructed. Phylogenetic trees were drawn using the Interactive Tree of Life (iTOL) programs [51].

3.9. Nucleotide Sequence Accession Number

The complete sequences of the chromosome of R31 and plasmid pR31-KPC were submitted to GenBank database, under accession numbers CP061850 and CP061851, respectively.

4. Conclusions

Carbapenems are still first-line agents for the treatment of *P. aeruginosa* infections. The occurrence and dissemination of bla_{KPC-2} and/or other carbapenemase gene-carrying *P. aeruginosa* strains have important clinical and epidemiological implications. Four other fully sequenced plasmids of the same incompatibility group as pR31-KPC have been reported in China, which suggest that it may be an important vehicle in the dissemination of bla_{KPC-2} in *Pseudomonas*.

Supplementary Materials: The following are available online at https://www.mdpi.com/article/10.3390/antibiotics10101234/s1. Figure S1: Schematic maps of p1011-KPC2, p14057A, YLH6_P3, pP23-KPC, and pR31-KPC. Figure S2: Evolutionary relationships of 202 *P. aeruginosa* genomes in the GenBank database. Table S1: Pairwise comparison of bla_{KPC-2}-carrying plasmids from *P. aeruginosa* using BLASTN. Table S2: Beta-lactamase and carbapenemase genes carried by *P. aeruginosa* genomes based on ResFinder results.

Author Contributions: Conceptualization, M.Y. and J.L.; methodology, M.Y., H.G., D.S. and W.C.; data analysis, M.Y., H.G., X.S., J.C. and J.L.; resources, H.G., D.S. and J.L.; writing—original draft, M.Y.; writing—review and editing, B.K. and J.L. All authors have read and agreed to the published version of the manuscript.

Funding: This work was supported by the National Natural Science Foundation of China (81861138053, 31761133004) and the Ministry of Science and Technology (2018ZX10733-402). None of the funders had any role in the study design, data collection, and interpretation or the decision to submit the work for publication.

Informed Consent Statement: Informed consent was obtained from all subjects involved in the study.

Data Availability Statement: The data presented in this study are available on request from the corresponding author. The plasmid sequences analyzed in this study can be found in public NCBI Genbank databases. The accession numbers were provided in this article when these plasmids were firstly indicated.

Conflicts of Interest: The authors declare no conflict of interest.

References

1. Gellatly, S.L.; Hancock, R.E. Pseudomonas aeruginosa: New insights into pathogenesis and host defenses. *Pathog. Dis.* **2013**, *67*, 159–173. [CrossRef] [PubMed]
2. Azam, M.W.; Khan, A.U. Updates on the Pathogenicity Status of Pseudomonas Aeruginosa. *Drug Discov. Today* **2019**, *24*, 350–359. [CrossRef] [PubMed]
3. Botelho, J.; Grosso, F.; Peixe, L. Antibiotic resistance in Pseudomonas aeruginosa—Mechanisms, epidemiology and evolution. *Drug Resist. Updates* **2019**, *44*, 100640. [CrossRef]
4. Breidenstein, E.B.; de la Fuente-Núñez, C.; Hancock, R.E. Pseudomonas aeruginosa: All roads lead to resistance. *Trends Microbiol.* **2011**, *19*, 419–426. [CrossRef] [PubMed]

5. Kung, V.L.; Ozer, E.A.; Hauser, A.R. The accessory genome of Pseudomonas aeruginosa. *Microbiol. Mol. Biol. Rev.* **2010**, *74*, 621–641. [CrossRef]
6. Cuzon, G.; Naas, T.; Villegas, M.V.; Correa, A.; Quinn, J.P.; Nordmann, P. Wide dissemination of Pseudomonas aeruginosa producing beta-lactamase bla$_{KPC-2}$ gene in Colombia. *Antimicrob. Agents Chemother.* **2011**, *55*, 5350–5353. [CrossRef] [PubMed]
7. Poirel, L.; Weldhagen, G.F.; Naas, T.; De Champs, C.; Dove, M.G.; Nordmann, P. *GES-2*, a class A beta-lactamase from Pseudomonas aeruginosa with increased hydrolysis of imipenem. *Antimicrob. Agents Chemother.* **2001**, *45*, 2598–2603. [CrossRef] [PubMed]
8. Senda, K.; Arakawa, Y.; Nakashima, K.; Ito, H.; Ichiyama, S.; Shimokata, K.; Kato, N.; Ohta, M. Multifocal outbreaks of metallo-beta-lactamase-producing Pseudomonas aeruginosa resistant to broad-spectrum beta-lactams, including carbapenems. *Antimicrob. Agents Chemother.* **1996**, *40*, 349–353. [CrossRef]
9. Lauretti, L.; Riccio, M.L.; Mazzariol, A.; Cornaglia, G.; Amicosante, G.; Fontana, R.; Rossolini, G.M. Cloning and characterization of blaVIM, a new integron-borne metallo-beta-lactamase gene from a Pseudomonas aeruginosa clinical isolate. *Antimicrob. Agents Chemother.* **1999**, *43*, 1584–1590. [CrossRef]
10. Murphy, T.A.; Simm, A.M.; Toleman, M.A.; Jones, R.N.; Walsh, T.R. Biochemical characterization of the acquired metallo-beta-lactamase SPM-1 from Pseudomonas aeruginosa. *Antimicrob. Agents Chemother.* **2003**, *47*, 582–587. [CrossRef] [PubMed]
11. Jovcic, B.; Lepsanovic, Z.; Suljagic, V.; Rackov, G.; Begovic, J.; Topisirovic, L.; Kojic, M. Emergence of NDM-1 metallo-beta-lactamase in Pseudomonas aeruginosa clinical isolates from Serbia. *Antimicrob. Agents Chemother.* **2011**, *55*, 3929–3931. [CrossRef] [PubMed]
12. Leiros, H.K.; Borra, P.S.; Brandsdal, B.O.; Edvardsen, K.S.; Spencer, J.; Walsh, T.R.; Samuelsen, O. Crystal structure of the mobile metallo-beta-lactamase AIM-1 from Pseudomonas aeruginosa: Insights into antibiotic binding and the role of Gln157. *Antimicrob. Agents Chemother.* **2012**, *56*, 4341–4353. [CrossRef] [PubMed]
13. Lee, W.; Chung, H.S.; Lee, Y.; Yong, D.; Jeong, S.H.; Lee, K.; Chong, Y. Comparison of matrix-assisted laser desorption ionization-time-of-flight mass spectrometry assay with conventional methods for detection of IMP-6, VIM-2, NDM-1, SIM-1, KPC-1, OXA-23, and OXA-51 carbapenemase-producing Acinetobacter spp., *Pseudomonas aeruginosa*, and *Klebsiella pneumoniae*. *Diagn. Microbiol. Infect. Dis.* **2013**, *77*, 227–230. [CrossRef] [PubMed]
14. Pollini, S.; Maradei, S.; Pecile, P.; Olivo, G.; Luzzaro, F.; Docquier, J.D.; Rossolini, G.M. FIM-1, a new acquired metallo-beta-lactamase from a Pseudomonas aeruginosa clinical isolate from Italy. *Antimicrob. Agents Chemother.* **2013**, *57*, 410–416. [CrossRef] [PubMed]
15. Pfennigwerth, N.; Lange, F.; Belmar Campos, C.; Hentschke, M.; Gatermann, S.G.; Kaase, M. Genetic and biochemical characterization of HMB-1, a novel subclass B1 metallo-beta-lactamase found in a Pseudomonas aeruginosa clinical isolate. *J. Antimicrob. Chemother.* **2017**, *72*, 1068–1073. [PubMed]
16. Boyd, D.A.; Lisboa, L.F.; Rennie, R.; Zhanel, G.G.; Dingle, T.C.; Mulvey, M.R. Identification of a novel metallo-beta-lactamase, CAM-1, in clinical Pseudomonas aeruginosa isolates from Canada. *J. Antimicrob. Chemother.* **2019**, *74*, 1563–1567. [CrossRef]
17. Sevillano, E.; Gallego, L.; Garcia-Lobo, J.M. First detection of the OXA-40 carbapenemase in P. aeruginosa isolates, located on a plasmid also found in *A. baumannii*. *Pathol. Biol.* **2009**, *57*, 493–495. [CrossRef]
18. Borah, V.V.; Saikia, K.K.; Hazarika, N.K. First report on the detection of OXA-48 beta-lactamase gene in Escherichia coli and Pseudomonas aeruginosa co-infection isolated from a patient in a Tertiary Care Hospital in Assam. *Indian J. Med. Microbiol.* **2016**, *34*, 252–253. [CrossRef]
19. Bonnin, R.A.; Bogaerts, P.; Girlich, D.; Huang, T.D.; Dortet, L.; Glupczynski, Y.; Naas, T. Molecular Characterization of OXA-198 Carbapenemase-Producing Pseudomonas aeruginosa Clinical Isolates. *Antimicrob. Agents Chemother.* **2018**, *62*, e02496-17. [CrossRef]
20. Yoon, E.J.; Jeong, S.H. Mobile Carbapenemase Genes in Pseudomonas aeruginosa. *Front. Microbiol.* **2021**, *12*, 614058. [CrossRef]
21. Naas, T.; Cuzon, G.; Villegas, M.V.; Lartigue, M.F.; Quinn, J.P.; Nordmann, P. Genetic structures at the origin of acquisition of the beta-lactamase bla$_{KPC}$ gene. *Antimicrob. Agents Chemother.* **2008**, *52*, 1257–1263. [CrossRef] [PubMed]
22. Akpaka, P.E.; Swanston, W.H.; Ihemere, H.N.; Correa, A.; Torres, J.A.; Tafur, J.D.; Montealegre, M.C.; Quinn, J.P.; Villegas, M.V. Emergence of KPC-producing Pseudomonas aeruginosa in Trinidad and Tobago. *J. Clin. Microbiol.* **2009**, *47*, 2670–2671. [CrossRef] [PubMed]
23. Poirel, L.; Nordmann, P.; Lagrutta, E.; Cleary, T.; Munoz-Price, L.S. Emergence of KPC-producing Pseudomonas aeruginosa in the United States. *Antimicrob. Agents Chemother.* **2010**, *54*, 3072. [CrossRef]
24. Robledo, I.E.; Aquino, E.E.; Vázquez, G.J. Detection of the KPC gene in Escherichia coli, Klebsiella pneumoniae, Pseudomonas aeruginosa, and Acinetobacter baumannii during a PCR-based nosocomial surveillance study in Puerto Rico. *Antimicrob. Agents Chemother.* **2011**, *55*, 2968–2970. [CrossRef] [PubMed]
25. Ge, C.; Wei, Z.; Jiang, Y.; Shen, P.; Yu, Y.; Li, L. Identification of KPC-2-producing Pseudomonas aeruginosa isolates in China. *J. Antimicrob. Chemother.* **2011**, *66*, 1184–1186. [CrossRef]
26. Jacome, P.R.; Alves, L.R.; Cabral, A.B.; Lopes, A.C.; Maciel, M.A. First report of KPC-producing Pseudomonas aeruginosa in Brazil. *Antimicrob. Agents Chemother.* **2012**, *56*, 4990. [CrossRef]
27. Hagemann, J.B.; Pfennigwerth, N.; Gatermann, S.G.; von Baum, H.; Essig, A. KPC-2 carbapenemase-producing Pseudomonas aeruginosa reaching Germany. *J. Antimicrob. Chemother.* **2018**, *73*, 1812–1814. [CrossRef]

28. Naas, T.; Bonnin, R.A.; Cuzon, G.; Villegas, M.V.; Nordmann, P. Complete sequence of two KPC-harbouring plasmids from Pseudomonas aeruginosa. *J. Antimicrob. Chemother.* **2013**, *68*, 1757–1762. [CrossRef]

29. Dai, X.; Zhou, D.; Xiong, W.; Feng, J.; Luo, W.; Luo, G.; Wang, H.; Sun, F.; Zhou, X. The IncP-6 Plasmid p10265-KPC from Pseudomonas aeruginosa Carries a Novel DeltaISEc33-Associated bla_{KPC-2} Gene Cluster. *Front. Microbiol.* **2016**, *7*, 310. [CrossRef]

30. Galetti, R.; Andrade, L.N.; Varani, A.M.; Darini, A.L.C. A Phage-Like Plasmid Carrying bla (KPC-2) Gene in Carbapenem-Resistant Pseudomonas aeruginosa. *Front. Microbiol.* **2019**, *10*, 572. [CrossRef]

31. Hu, Y.Y.; Wang, Q.; Sun, Q.L.; Chen, G.X.; Zhang, R. A novel plasmid carrying carbapenem-resistant gene bla(KPC-2) in Pseudomonas aeruginosa. *Infect. Drug Resist.* **2019**, *12*, 1285–1288. [CrossRef] [PubMed]

32. Shi, L.; Liang, Q.; Feng, J.; Zhan, Z.; Zhao, Y.; Yang, W.; Yang, H.; Chen, Y.; Huang, M.; Tong, Y.; et al. Coexistence of two novel resistance plasmids, bla(KPC-2)-arrying p14057A and tetA(A)-carrying p14057B, in Pseudomonas aeruginosa. *Virulence* **2018**, *9*, 306–311. [CrossRef]

33. Sun, F.; Zhou, D.; Wang, Q.; Feng, J.; Feng, W.; Luo, W.; Liu, Y.; Qiu, X.; Yin, Z.; Xia, P. Genetic characterization of a novel blaDIM-2-carrying megaplasmid p12969-DIM from clinical Pseudomonas putida. *J. Antimicrob. Chemother.* **2016**, *71*, 909–912. [CrossRef]

34. Xiong, J.; Alexander, D.C.; Ma, J.H.; Deraspe, M.; Low, D.E.; Jamieson, F.B.; Roy, P.H. Complete sequence of pOZ176, a 500-kilobase IncP-2 plasmid encoding IMP-9-mediated carbapenem resistance, from outbreak isolate Pseudomonas aeruginosa 96. *Antimicrob. Agents Chemother.* **2013**, *57*, 3775–3782. [CrossRef] [PubMed]

35. Roy Chowdhury, P.; Merlino, J.; Labbate, M.; Cheong, E.Y.; Gottlieb, T.; Stokes, H.W. Tn6060, a transposon from a genomic island in a Pseudomonas aeruginosa clinical isolate that includes two class 1 integrons. *Antimicrob. Agents Chemother.* **2009**, *53*, 5294–5296. [CrossRef] [PubMed]

36. Di Pilato, V.; Pollini, S.; Rossolini, G.M. Tn6249, a new Tn6162 transposon derivative carrying a double-integron platform and involved with acquisition of the bla_{VIM-1} metallo-beta-lactamase gene in Pseudomonas aeruginosa. *Antimicrob. Agents Chemother.* **2015**, *59*, 1583–1587. [CrossRef]

37. Coyne, S.; Courvalin, P.; Galimand, M. Acquisition of multidrug resistance transposon Tn6061 and IS6100-mediated large chromosomal inversions in Pseudomonas aeruginosa clinical isolates. *Microbiology* **2010**, *156*, 1448–1458. [CrossRef] [PubMed]

38. Toleman, M.A.; Walsh, T.R. Combinatorial events of insertion sequences and ICE in Gram-negative bacteria. *FEMS Microbiol. Rev.* **2011**, *35*, 912–935. [CrossRef] [PubMed]

39. Aziz, R.K.; Bartels, D.; Best, A.A.; DeJongh, M.; Disz, T.; Edwards, R.A.; Formsma, K.; Gerdes, S.; Glass, E.M.; Kubal, M.; et al. The RAST Server: Rapid annotations using subsystems technology. *BMC Genom.* **2008**, *9*, 75. [CrossRef] [PubMed]

40. Boratyn, G.M.; Camacho, C.; Cooper, P.S.; Coulouris, G.; Fong, A.; Ma, N.; Madden, T.L.; Matten, W.T.; McGinnis, S.D.; Merezhuk, Y.; et al. BLAST: A more efficient report with usability improvements. *Nucleic Acids Res.* **2013**, *41*, W29–W33. [CrossRef] [PubMed]

41. Boutet, E.; Lieberherr, D.; Tognolli, M.; Schneider, M.; Bansal, P.; Bridge, A.J.; Poux, S.; Bougueleret, L.; Xenarios, I. UniProtKB/Swiss-Prot., the Manually Annotated Section of the UniProt KnowledgeBase: How to Use the Entry View. *Methods Mol. Biol.* **2016**, *1374*, 23–54. [PubMed]

42. O'Leary, N.A.; Wright, M.W.; Brister, J.R.; Ciufo, S.; Haddad, D.; McVeigh, R.; Rajput, B.; Robbertse, B.; Smith-White, B.; Ako-Adjei, D.; et al. Reference sequence (RefSeq) database at NCBI: Current status, taxonomic expansion, and functional annotation. *Nucleic Acids Res.* **2016**, *44*, D733–D745. [CrossRef] [PubMed]

43. Jia, B.; Raphenya, A.R.; Alcock, B.; Waglechner, N.; Guo, P.; Tsang, K.K.; Lago, B.A.; Dave, B.M.; Pereira, S.; Sharma, A.N.; et al. CARD 2017: Expansion and model-centric curation of the comprehensive antibiotic resistance database. *Nucleic Acids Res.* **2017**, *45*, D566–D573. [CrossRef]

44. Zankari, E.; Hasman, H.; Cosentino, S.; Vestergaard, M.; Rasmussen, S.; Lund, O.; Aarestrup, F.M.; Larsen, M.V. Identification of acquired antimicrobial resistance genes. *J. Antimicrob. Chemother.* **2012**, *67*, 2640–2644. [CrossRef] [PubMed]

45. Siguier, P.; Varani, A.; Perochon, J.; Chandler, M. Exploring bacterial insertion sequences with ISfinder: Objectives, uses, and future developments. *Methods Mol. Biol.* **2012**, *859*, 91–103.

46. Varani, A.M.; Siguier, P.; Gourbeyre, E.; Charneau, V.; Chandler, M. ISsaga is an ensemble of web-based methods for high throughput identification and semi-automatic annotation of insertion sequences in prokaryotic genomes. *Genome Biol.* **2011**, *12*, R30. [CrossRef]

47. Moura, A.; Soares, M.; Pereira, C.; Leitao, N.; Henriques, I.; Correia, A. INTEGRALL: A database and search engine for integrons, integrases and gene cassettes. *Bioinformatics* **2009**, *25*, 1096–1098. [CrossRef]

48. Tansirichaiya, S.; Rahman, M.A.; Roberts, A.P. The Transposon Registry. *Mob. DNA* **2019**, *10*, 40. [CrossRef]

49. Edgar, R.C. MUSCLE: Multiple sequence alignment with high accuracy and high throughput. *Nucleic Acids Res.* **2004**, *32*, 1792–1797. [CrossRef]

50. Gardner, S.N.; Slezak, T.; Hall, B.G. kSNP3.0: SNP detection and phylogenetic analysis of genomes without genome alignment or reference genome. *Bioinformatics* **2015**, *31*, 2877–2878. [CrossRef]

51. Letunic, I.; Bork, P. Interactive Tree of Life (iTOL) v4: Recent updates and new developments. *Nucleic Acids Res.* **2019**, *47*, W256–W259. [CrossRef] [PubMed]

Article

Diversity of International High-Risk Clones of *Acinetobacter baumannii* Revealed in a Russian Multidisciplinary Medical Center during 2017–2019

Andrey Shelenkov [1,*], Lyudmila Petrova [2], Mikhail Zamyatin [2], Yulia Mikhaylova [1] and Vasiliy Akimkin [1]

[1]   Central Research Institute of Epidemiology, 111123 Moscow, Russia; mihailova@cmd.su (Y.M.); vgakimkin@yandex.ru (V.A.)

[2]   National Medical and Surgical Center, 105203 Moscow, Russia; lutix85@yandex.ru (L.P.); mnz1@yandex.ru (M.Z.)

*   Correspondence: fallandar@gmail.com

Abstract: *Acinetobacter baumannii* is a dangerous bacterial pathogen possessing the ability to persist on various surfaces, especially in clinical settings, and to rapidly acquire the resistance to a broad spectrum of antibiotics. Thus, the epidemiological surveillance of *A. baumannii* within a particular hospital, region, and across the world is an important healthcare task that currently usually includes performing whole-genome sequencing (WGS) of representative isolates. During the past years, the dissemination of *A. baumannii* across the world was mainly driven by the strains belonging to two major groups called the global clones or international clones (ICs) of high risk (IC1 and IC2). However, currently nine ICs are already considered. Although some clones were previously thought to spread in particular regions of the world, in recent years this is usually not the case. In this study, we determined five ICs, as well as three isolates not belonging to the major ICs, in one multidisciplinary medical center within the period 2017–2019. We performed WGS using both short- and long-read sequencing technologies of nine representative clinical *A. baumannii* isolates, which allowed us to determine the antibiotic resistance and virulence genomic determinants, reveal the CRISPR/Cas systems, and obtain the plasmid structures. The phenotypic and genotypic antibiotic resistance profiles are compared, and the possible ways of isolate and resistance spreading are discussed. We believe that the data obtained will provide a better understanding of the spreading and resistance acquisition of the ICs of *A. baumannii* and further stress the necessity for continuous genomic epidemiology surveillance of this problem-causing bacterial species.

Keywords: *Acinetobacter baumannii*; antibiotic resistance; virulence; whole-genome sequencing; international high-risk clones; genomic epidemiology

Citation: Shelenkov, A.; Petrova, L.; Zamyatin, M.; Mikhaylova, Y.; Akimkin, V. Diversity of International High-Risk Clones of *Acinetobacter baumannii* Revealed in a Russian Multidisciplinary Medical Center during 2017–2019. *Antibiotics* **2021**, *10*, 1009. https://doi.org/10.3390/antibiotics10081009

Academic Editor: Manuela Oliveira

Received: 23 July 2021
Accepted: 18 August 2021
Published: 20 August 2021

Publisher's Note: MDPI stays neutral with regard to jurisdictional claims in published maps and institutional affiliations.

1. Introduction

Monitoring the spread of particular lineages of pathogenic bacteria and the associated antimicrobial resistance determinants within a particular hospital, country, or across the world represents a very important task for national and international public health institutions. Currently, such surveillance is becoming more and more dependent on next-generation sequencing (NGS) of bacterial genomes and the corresponding bioinformatics analysis pipelines [1–3]. Such investigations have already formed a new field called 'genomic epidemiology', in which methods allow to obtain huge amounts of epidemiologically and medically relevant information in a cost-, time-, and resource-efficient way [4–6].

During recent years, the antibiotic resistance within different species of nosocomial and community-acquired bacterial pathogens has increased to dangerous levels [7–9]. This problem cannot be solved without comprehensive investigations of resistance transmission mechanisms and global epidemiological surveillance.

Currently, one of the most problematic pathogens is *Acinetobacter baumannii*, accounting for about 2% of all healthcare-associated infections in USA and Europe [10] and up to 4% in Asia [11]. *A. baumannii* is a member of the ESKAPE group of bacterial pathogens, also including *Enterococcus faecium*, *Staphylococcus aureus*, *Klebsiella pneumoniae*, *Pseudomonas aeruginosa*, and *Enterobacter* spp., which are major causes of antibiotic-resistant infections worldwide [12]. It is a Gram-negative coccobacillus that is mainly responsible for causing pneumonia and wound infections associated with elevated morbidity and mortality in clinical settings [13]. The notable characteristics of *A. baumannii* include the ability to rapidly acquire multidrug-, extensive drug-, and even pandrug-resistance phenotypes [14], as well as to easily survive and transfer in the hospital environment, such as attaching to various biotic and abiotic surfaces [15]. *A. baumannii* evolution during the past five decades was mainly driven by two globally disseminated clones, GC1 and GC2 (also called IC1 and IC2, IC standing for 'international clone') [16]. However, six additional clonal lineages are currently generally accepted [17], and IC9 is on its way [18].

In Russia, *A. baumannii* constitutes up to 16.8% of healthcare-associated bacterial infections and exhibits a high rate of carbapenem resistance (about 77%), with the predominating clones being IC1, IC2, and IC6 [19].

Although many reports consider different clones or lineages of *A. baumannii* species to be associated with particular parts or regions of the world [13,20–22], the dramatically fast distribution of SARS-CoV-2 in 2020 has demonstrated that our knowledge regarding the spread of different pathogens is still limited.

Here we present the results of genomic epidemiology monitoring of *A. baumannii* in a multidisciplinary medical center in Moscow, Russia, during the period 2017–2019. Amazingly, we have revealed the isolates belonging to 5 out of 9 international clonal lineages (ICL), as well as additional isolates not clustering to any known ICL within our samples. We selected nine representative isolates for this manuscript and performed whole-genome sequencing for them using second- and third-generation (long-read) sequencing technologies. Comprehensive analyses of phenotypic and genotypic antimicrobial resistance, virulence factors, plasmids, and CRISPR arrays for the selected isolates are provided.

We believe that our data will facilitate a better understanding of *A. baumannii* spread across the world and the possible ways of acquiring antimicrobial resistance by this dangerous pathogen.

2. Results

2.1. Isolate Metadata and Typing

The metadata for the isolates and the results of their typing using the Pasteur MLST scheme, K-, and OCL-loci are presented in Table 1.

Table 1. The origin and typing of the clinical *A. baumannii* isolates studied.

Sample id	Patient Code	Isolation Date	Clinical Department	Locus	MLST	OCL-Type	KL-Type	IC
CriePir33	P1	03.05.2017	Traumatology	Wound	ST78	OCL1	KL3	IC6 (CC78)
CriePir87	P2	04.07.2017	Surgery	Soft tissue abscess	ST2	OCL1	KL33	IC2 (CC2)
CriePir168	P3	03.12.2017	ICU	Urine	ST1	OCL1	KL17	IC1 (CC1)
CriePir254	P4	24.08.2018	Surgery	Bile	ST370	OCL1	KL25	CC252
CriePir298	P5	10.09.2019	CNS Rehabilitation	Urine	ST25	OCL6	KL116	IC7 (CC25)
CriePir306	P6	05.08.2019	ICU	BAL	ST15	OCL7	KL9	IC4 (CC15)
CriePir307	P7	29.06.2019	ICU	BAL	ST1487 *	OCL2	KL45	CC152
CriePir308	P6	22.08.2019	ICU	CVC	ST2	OCL7	KL9	IC2 (CC2)
CriePir309	P8	31.08.2019	CNS Rehabilitation	Urine	ST911	OCL6	KL14	CC132

ICU—intensive care unit; CVC—central venous catheter; BAL—bronchoalveolar lavage; *—novel ST identified by us.

In this study, we aimed to capture the widest possible diversity of *A. baumannii* clones while keeping the number of isolates lower for the sake of presentation clarity. We selected nine isolates, six of which represented known international clones of high risk and three isolates represented singletons that could not be attributed to any ICs. Well-established 'old' international clones 1 and 2 [13] are presented by their general sequence types. Two isolates representing IC2 had different K- and OCL-types, so they were both included in the study to further increase the isolate diversity.

IC4 and IC6 were also represented by their major STs [17]. It is interesting that the isolates representing two different clones, CriePir306 (IC4) and CriePir308 (IC2), were isolated from the same patient, but from different sites. However, they had the same K- and OCL-types, which is rather surprising.

CriePir307 possessed a novel sequence type that was not clustered with existing international clones by either cgMLST or MLST loci. Although it possessed the $bla_{\text{OXA-64}}$ gene that is known to be the characteristic of IC7 [23], it was quite different from other IC7 isolates by cgMLST profile (see Table S1 and Figure S1). Thus, it was assigned to the most frequent Pasteur ST profile clustering in the same clonal complex (CC152). CriePir298, which belonged to IC7, possessed a general MLST sequence type for this clone—ST25 [17]. However, it had a rather rare capsular type—KL116 [24].

CriePir254 and CriePir309 also did not belong to the known international clonal lineages, but could be attributed to CC252 and CC132, respectively.

The minimum spanning tree for the isolates is presented in Figure 1. It is not surprising that the isolates are located very far from each other in terms of their cgMLST profiles since the aim of this study was to capture the maximal possible diversity. Only the isolates CriePir87 and CriePir308 belonging to IC2 were comparatively close to each other (198 allele differences).

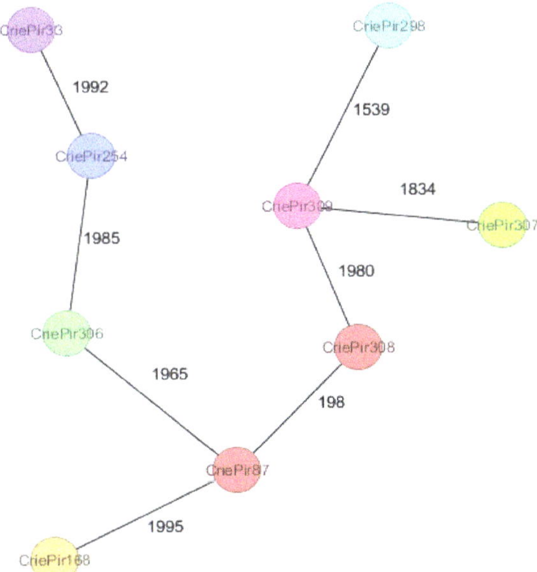

Figure 1. The minimum spanning tree for the isolates studied based on cgMLST profiles. The numbers indicate the amount of different alleles between the pairs of corresponding isolates. Close isolates (CriePi87 and CriePi308) are shown in the same color; other isolates are not close to each other.

We also built a cgMLST tree for our isolates and the whole set of RefSeq isolates (accessed on 20 March 2021), trying to infer the possible spreading information. A subset of this tree, including the closest matches from RefSeq to our isolates in terms of the number of cgMLST allele differences, is provided in the Figure S1.

As we can see from this tree, the isolates with unusual profiles, such as CriePir254 (rare ST, no IC) and CriePir307 (novel ST, no IC), do not have close relatives in terms of the number of allele differences. Moderately close neighbors were revealed for CriePir168 (IC1, GCA_000830055.1, Australia, distance = 187), CriePir308 (IC2, GCA_000314655.1, USA, distance = 50), CriePir306 (IC4, GCF_003583665.1, Spain, distance = 147), and CriePir33 (IC6, GCF_003948375.1, USA, distance = 87). However, this information does not allow inferring the spreading routes of the isolates, and more genomic data is required to achieve this goal.

Complete cgMLST profiles for all our isolates and their closest matches from RefSeq are presented in Table S1.

2.2. Antimicrobial Resistance

Phenotypic characterization and the genotypic resistance determinants of the isolates studied are presented in Figure 2.

Sample id / Antibiotics and resistance genes	ARR3	ant(3)-IIa	ant(3)-Ia	aac(3)-Ia	aac(3)-IIa	aac(6')-Iaf	aac(6')-Ib-cr	aph(3')-Ia	aph(3')-VIa	aph(3'')-VIb	armA	blaADC	blaCARB-16	blaOXA-51-like	blaOXA-23	blaOXA-72	blaPER	catA1	catB8	cmlA1	floR	mphE	msrE	sul1	sul2	tet(B)	Amikacin	Gentamicin	Tobramycin	Imipenem	Meropenem	Levofloxacin	Ciprofloxacin	Tmp/Smz
CriePir33	-	+	-	-	-	-	-	-	-	-	+	6	+	90	-	+	-	-	-	-	+	+	+	+	+	-	R	R	R	R	R	R	R	R
CriePir87	+	+	+	-	-	-	+	+	-	-	+	30	-	66	+	-	7	-	+	+	-	+	+	+	+	-	R	R	R	R	R	R	R	R
CriePir168	-	+	-	-	-	-	-	-	-	-	-	74	-	69	-	+	-	-	-	-	-	-	-	-	-	-	S	R	S	R	R	R	R	S
CriePir254	-	+	-	-	-	-	-	-	-	-	-	79	-	98	-	-	-	-	-	-	-	-	-	-	-	-	S	S	S	S	S	S	S	S
CriePir298	+	+	-	-	+	-	-	-	-	-	+	79	-	64	+	-	7	-	-	+	-	+	+	+	+	+	R	R	R	R	R	R	R	R
CriePir306	-	+	-	-	-	-	-	-	+	-	-	6	-	51	-	-	-	-	-	-	-	-	-	-	+	-	R	R	S	S	S	R	R	S
CriePir307	-	+	-	-	-	-	-	-	-	-	-	3	-	64	-	-	-	-	-	-	-	-	-	-	-	-	S	S	S	S	S	S	S	S
CriePir308	-	+	-	+	-	+	-	-	-	-	-	30	-	66	-	+	-	-	-	-	-	-	-	-	-	+	R	R	S	R	R	R	R	S
CriePir309	-	+	-	-	-	-	-	-	-	-	-	2	-	120	-	-	-	-	-	-	-	-	-	-	-	-	S	S	S	S	S	S	S	S

Figure 2. Antimicrobial resistance of the *A. baumannii* isolates studied. Corresponding antibiotics and resistance genes are filled with the same colors. The numbers for the *bla* genes indicate the corresponding variants for the sake of brevity. Tmp/Smz—trimethoprim/sulfamethoxazole.

As we can see from Figure 2, the isolates form three groups containing three members each: multidrug-resistant (CriePir33, CriePir87, and CriePir298), which exhibit resistance to all antibiotics from the panel; non-resistant (CriePir254, CriePir307, and CriePir309), possessing only intrinsic oxacillinase genes); and intermediate (CriePir168, CriePir306, and CriePir308), having resistance only to some of the antibiotics tested. Five isolates were carbapenem-resistant, and the likely mechanism of such a resistance is expression of $bla_{OXA\text{-}23}$ (CriePir87 and CriePir298) and $bla_{OXA\text{-}72}$ (CriePir33, CriePir168, and CriePir308) carbapenemase genes. In addition, the $bla_{OXA\text{-}23}$-carrying isolates also encoded PER-7 extended spectrum β-lactamase (ESBL), which was reported to demonstrate high activity against broad-spectrum cephalosporins in *A. baumannii* [25].

The resistant isolates also included genes and gene clusters providing resistance to aminoglycosides (for example, *armA*), sulfamethoxazole (*sul1* and *sul2*), chloramphenicol (*cmlA1*), and tetracycline (*tet(B)*). However, the latter two antimicrobials were not included in the panel.

Interestingly, CriePir33 possessed the $bla_{CARB\text{-}16}$ gene representing a rather rarely occurring $bla_{CARB\text{-}5}$-like class A beta-lactamase gene, which was first revealed in *Acinetobacter pittii* [26]. The enzyme encoded by $bla_{CARB\text{-}16}$ differs only by one amino acid substitution from $bla_{CARB\text{-}5}$.

In general, the isolates demonstrated very good compliance between the phenotypic and genotypic characteristics of their antimicrobial resistance. The group of susceptible isolates included only intrinsic blaADC and blaOXA-51-like oxacillinases, as well as the *ant(3)-IIa* gene, while the isolates that exhibited the multidrug-resistant phenotype possessed the largest number of acquired resistance genes (8 for CriePir33, 13 for CriePir87, and 11 for CriePir298, respectively).

2.3. Virulence Genes

The distribution of the virulence genes in the isolates studied is shown in Figure 3.

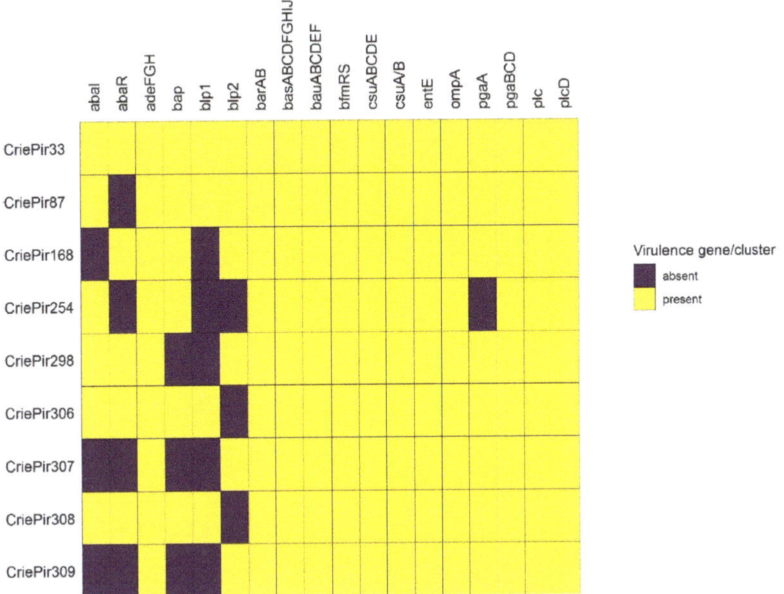

Figure 3. Virulence factors of the *A. baumannii* isolates studied. Genes constituting the same cluster presented in all isolates were combined for the sake of brevity. abaI, abaR—components of the quorum sensing system; adeFGH—efflux pump; bap—biofilm-associated protein; blp1,blp2—bap-like proteins; barAB—siderophore efflux system of the ABC superfamily; basABCDFGHIJ—proteins involved in biosynthesis of acinetobactin; bauABCDEF—receptor for ferric-acinetobactin complexes; bfmRS—two-component signal transduction system; csuA/BABCDE—Csu pili; entE—enterobactin biosynthesis; ompA—outer membrane protein; pgaABCD—biofilm formation locus; plc, plcD—phospholipase genes.

The virulence gene sets of all isolates were quite similar. They mainly included the factors involved in biofilm formation (*adeFGH*, *csu*, and *pga* clusters) as well as the *bau* and *bas* clusters involved in the iron acquisition system and acinetobactin functioning.

CriePir306 and CriePir308 included all 39 factors revealed, while the other isolates lacked from one to four virulence genes each. Interestingly, we have not revealed any virulence plasmids, and all virulence genes were located on chromosomes.

The list of NCBI accession numbers for the virulence genes revealed is shown in Table S2.

2.4. Plasmids

The isolates had from one to six plasmids each (Table S3). However, these plasmids usually did not carry the resistance genes, except for CriePir298, which possessed the plasmid containing all resistance determinants, excluding *bla*$_{OXA-23}$ and the intrinsic genes, as well as CriePir33, CriePir168, and CriePir308, the latter two of which carried two copies of *bla*$_{OXA-72}$ of plasmid origin. Interestingly, the plasmids of CriePir168 (length = 10,878 bp) and CriePir308 (length = 10,879 bp) had essentially the same sequences except for several deletions that might result from long-read assembly algorithm imperfections. These two isolates belonged to different clonal lineages (ST1 and ST2, respectively) and were isolated with an interval of 2 years. However, both of them were found in the same clinical department (ICU), and thus the persistence of the bla_{OXA-72}-carrying plasmid within this unit could be suggested. Another interesting fact was that these plasmids had 99.7% identity with the pAB120 plasmid (Genbank accession CP031446.1) of the MDR-UNC *A. baumannii* isolate (ST2), which caused a fatal case of necrotizing fasciitis in a USA hospital in 2019 [27].

CriePir298 included three plasmids, the smallest of which (length = 2343) had exactly the same sequence as pA85-1a (Genbank accession CP021784.1) from the A85 *A. baumannii* strain isolated in Australia in 2003. This plasmid was also found in several other isolates of IC1 [28], but its functional properties were not described. However, CriePir298 belonged to IC7. A brief description of the plasmids is provided in Table 2.

Table 2. Plasmid characteristics of the isolates studied.

Sample Id	Number of Plasmids	Plasmid Sizes	Virulence/Resistance Determinants in Plasmids	Related Plasmids from Other Isolates
CriePir33	1	17,765	bla_{OXA-72}	pIBAC_oxa58_20C15, KY202458.1
CriePir87	1	11,194	-	pUnnamed1, CP035673.1
CriePir168	3	1217–10,878	bla_{OXA-72}	pAB120, CP031446.1
CriePir254	2	10,427–90,326	-	pGFJ6, CP016902.1
CriePir298	3	2343–183,139	*mph(E), msr(E), armA, sul1, sul2, blaPER-7, ARR-3, aph(6)-Id, aph(3″)-Ib, aac(3)-IIa, tet(B)*	pA85-1a, CP021784.1
CriePir306	4	2845–80,829	-	pA1296_1, CP018333.1
CriePir307	6	2278–114,430	-	pABAY15001_6E, MK386684.1
CriePir308	1	10,879	bla_{OXA-72}	pAB120, CP031446.1
CriePir309	2	11,195–94,551	-	pTS134338, CP042210.1

2.5. CRISPR Arrays and CRISPR/Cas Systems

CRISPR arrays were revealed in seven out of nine isolates (except CriePir87 and CriePir308). However, at least five repeats with an evidence level = 4 were found only in four isolates (CriePir168, CriePir298, CriePir307, and CriePir309), and all of them except the latter possessed a full CAS-Type IF system. Interestingly, these CAS systems had different subtypes, namely, IF-1 for CriePir168 and CriePir298, and IF-2 for CriePir307. Detailed analysis of the differences between these two subtypes lies beyond the scope of this manuscript.

Interestingly, although CriePir309 had a highly confident CRISPR array with 17 repeats, it was also the only isolate carrying an anti-CRISPR element, which could prevent the development of a functional CRISPR/Cas system.

Most repeats and all *cas* genes were located on the chromosomes. CriePir168 had one array on the short plasmid (length = 2372), and CriePir308 had one of seven arrays on another plasmid (length = 14,128). However, both of these repeats had a low evidence level.

Detailed characteristics of the CRISPR elements, including the chromosome positions, repeat sequences, and accession numbers for the *cas* genes revealed, are provided in Table S4.

3. Discussion

In this manuscript, we presented the genomic epidemiology investigation of a diverse *A. baumannii* population within a multidisciplinary medical center in Moscow, Russia, for the period 2017–2019. We carefully selected the isolates representing the unusual international clone variability, as well as additional isolates belonging to singleton clones, and performed their long-read sequencing in order to obtain highly accurate chromosome sequences and delineate plasmids. Hybrid short- and long-read assemblies allowed us to improve the prediction of virulence and the resistance genomic determinants, as well as to retrieve additional information required for application of genomic epidemiology tools such as cgMLST analysis.

In order to perform the epidemiological surveillance and track the spreading of pathogenic bacteria within a particular healthcare facility, some geographical region, or across the world, a reliable typing scheme based on molecular or genomic characteristics of the isolates is required. Many typing schemes are already provided for pathogenic bacteria, including *A. baumannii*. Exemplary schemes/profiles comprise the ones based on the nucleotide frequency matrices for genomic sequences [29], CRISPR sequences [30], multilocus sequence typing (MLST) [31,32] based on seven housekeeping genes, MLST/KL loci [33], and core genome MLST (cgMLST) based on 2390 genes [34]. The latter was recently successfully applied for an outbreak investigation [35]. In addition, intrinsic carbapenemase *bla*$_{OXA-51}$-like gene variants were also proposed as a tool for *A. baumannii* identification and typing [36].

However, one of the most commonly used concepts for molecular epidemiology investigations is epidemic clonal lineages, or simply 'international clones', which represent genetically distinct populations of *A. baumannii* successfully spreading in different geographic locations [37]. Eight international clones were defined earlier [17], and the ninth was introduced recently [18]. The first three clones (IC1–IC3) are distributed worldwide, with IC1 and IC2 also known as global clones (GC), while the rest were sometimes defined as regional, or even endemic clones [13,20–22].

In contrast to this, we revealed an isolate belonging to IC7 in our hospital, although this clone was recently reported to be prevalent in South America and not in Europe [22]. Although our initial set of isolates contained less than 50 sequenced samples, we managed to reveal five different ICs and three singleton isolates not attributed to known clonal lineages. One of these isolates belonged to IC7: CriePir298 possessed the founder sequence type ST25, the members of which were revealed in Bolivia [22]. It is interesting that the ST1487 possessed by CriePir307 is rather different from ST25 in its allelic profile, and they do not cluster according to the eBURST analysis; at the same time, these two isolates share the same intrinsic OXA-51 variant, namely, OXA-64, and thus likely belong to the same lineage IC7 [23]. However, a large number of MLST and cgMLST allele differences did not allow to assign CriePir307 to IC7, and it was closer to the members of CC152. Another intriguing fact is that the resistance characteristics of CriePir298 and CriePir307 are completely different—the former was resistant to all antibiotics from the panel, while the latter was susceptible to all antibiotics tested. However, genomic analysis revealed that most resistance genes of CriePir298 were located on the resistance plasmid, which was very similar to pMC1.1 (Genbank accession MK531536.1) revealed in Bolivia in the investigation mentioned above [22].

Historic global clones were represented in our set by IC1 (CriePir168) and IC2 (CriePir87 and CriePir308). IC4, possessing the *bla*$_{OXA-51}$ intrinsic gene, was revealed in CriePir306 (ST15). This clone was previously found in South America [38], but it is also present in Europe [39]. IC6, which was once considered to be a Russian endemic [20] and constituted about 25% of clinical isolates in Russia in 2015–2016 [19], was presented by CriePir33, a carbapenem-resistant isolate carrying, among others, *bla*$_{CARB-16}$ and *bla*$_{OXA-72}$ resistance determinants. This complies with the observation that IC6 isolates from Russia usually obtain carbapenem resistance by acquiring class D carbapenemase genes and not by other mechanisms such as intrinsic gene mutations [19]. Unfortunately, the cgMLST comparison with the reference isolates did not reveal any clues regarding the routes of their spreading across the world due to the ubiquitous presence of widespread ICs in various geographical regions.

Finally, two isolates, CriePir254 and CriePir309, could not be attributed to known international clones. CriePir309, carrying *bla*$_{OXA-120}$ intrinsic beta-lactamase, can be assigned to the known complex CC132, although it is also rather close to CC33 [40], while CriePir254 is close to CC252. CriePir309 has a very rare ST911; currently, no full genomic sequences for the isolates of this sequence type are available in Genbank, and there is no information, except a definition, for this ST in the PubMLST database (https://pubmlst.org, accessed in 23 April 2021). This isolate was obtained from the only patient involved in the study living outside the Moscow region, and it possessed various interesting properties. For example, it did not carry resistance genes except intrinsic variants of the bla_{ADC} and bla_{OXA-51}-like genes, and it was the only isolate having anti-CRISPR proteins encoded (AcrIF11). At the same time, it carried several virulence gene clusters (*adeFGH, basABCDFGHIJ, csuABCD,* but not *bap*), which made it more similar to the strains causing community-acquired infections. CriePir254 and CriePir307 were similar to CriePir309 in this sense, since they also did not have resistance determinants. Thus, we can conclude that the isolates from our dataset that appeared to be rather distant from the known international high-risk clones were in fact less dangerous in terms of their antibiotic resistance.

In summary, our results confirm that in the era of globalization and rapid pathogen spread across the world, the concept of endemic clones becomes obsolete. For example, we revealed IC7 in Russia, and IC6, which was previously attributed to Europe, was recently found in Brazil [41]. In addition, a recent publication of the spatio-temporal distribution of *A. baumannii* in Germany showed that the isolates belonging to IC1, 2, 4, 6, and 7 were revealed during 2000–2018 [39], which corresponds to our dataset. However, the remarkable feature of our study is the discovery of such an IC variety within one hospital during a limited period of time, and for the patients living in the Moscow region and without a history of international travelling shortly before their hospital admission. This finding will allow developing updated prevention strategies and epidemiological measures to limit further high-risk clone spreading.

While epidemiological data is vitally important for studying the spread of any pathogenic bacteria, the information regarding the presence of antibiotic resistance and virulence factors in the isolates studied, as well as the possible mechanisms of their transfer, is no less important for developing the prevention measures. Sequencing on MinION and hybrid short-long read assembly allowed us to identify the locations of genes encoding oxacillinases and other resistance genes. The bla_{OXA-23} genes of two isolates (CriePir87 and CriePir298) were located on the chromosome, while bla_{OXA-72} for each of the three isolates (CriePir33, CriePir168, and CriePir308) were revealed on plasmids. Such a distribution complies well with previous studies [36,42].

In general, the antibiotic resistance genes of our isolates were not located on plasmids, except for CriePir298 and the three isolates mentioned in the previous paragraph. However, plasmid investigation can contribute not only to antimicrobial resistance studies but also to tracing the spread of the pathogens across the world, although the plasmid complement is not a reliable measure of relatedness [28]. For example, in our isolates we revealed the plasmids identical to the ones of clinical isolates from the USA and Australia. However, these isolates belonged to different clonal lineages than ours, so the possible ways of plasmid transferring between them cannot be easily reconstructed.

In contrast, we have not revealed any virulence plasmids in our isolates, and all the virulence genes were located on the chromosomes.

Virulence factors were represented in all of the isolates by the members of *csu* and *pga* clusters involved in biofilm formation [43,44], as well as the members of *bau* and *bas* clusters taking part in the iron acquisition system and acinetobactin transport and biosynthesis, respectively [45,46]. In addition, all isolates included the genes of the *adeFGH* efflux pump, the overexpression of which was also found to be associated with biofilm formation [47]. These genes are rather common for clinical *A. baumannii* isolates and, together with the other genes revealed (e.g., *bap* for CriePir87 and CriePir306), provide the environmental persistence for them [14]. In addition, all isolates contained the *ompA* gene, encoding a major component of outer membrane vesicles, which was considered to be a crucial virulence factor of *A. baumannii* [48]. Interestingly, the *bap* gene was revealed only in the isolates lacking the CRISPR/Cas system; this could be the result of preventing horizontal gene transfer by this system.

The sets of virulence factors were similar for all isolates, except for *abaR*, which was not revealed in the fully susceptible isolates (CriePir254, CriePir307, and CriePir309) and CriePir87, and *abaI*, which was not found in CriePir168, CriePir307, and CriePir309. These genes are involved in quorum sensing and may contribute to motility and host–pathogen interaction [45]. The presence of *abaI/abaR* was positively correlated with bacterial resistance rates [49], and thus their absence in susceptible isolates complies with this finding. However, additional investigations are needed to elucidate the possible mechanisms since such a correlation was not perfect for our isolates.

Finally, we can conclude that obtaining extensive data on the spreading of particular strains, high-risk clones, antimicrobial resistance, and virulence factors across a particular hospital, country, and region greatly facilitates developing the epidemiological measures for preventing an exponential increase in MDR *A. baumannii* strains, as well as other pathogenic bacterial species. Unfortunately, these measures are not sufficient for fighting the resistant bacteria in clinical settings. Possible holistic approaches to cope with this problem could include antibiotic stewardship [50], developing novel antibiotics [51], using bacteriophages, and other antibacterial moieties, such as antibodies, synthetic membrane-active agents, or antimicrobial peptides [52–55].

4. Materials and Methods

4.1. Determination of Antibiotic Susceptibility

Species identification for all isolates was performed by time-of-flight mass spectrometry (MALDI-TOF MS) using the VITEK MS system (bioMerieux, Marcy-l'Étoile, France), and the susceptibility to antimicrobials was determined by the disc diffusion method using the Mueller–Hinton medium (bioMerieux, Marcy-l'Étoile, France) and disks with antibiotics (BioRad, Marnes-la-Coquette, France), and by the Minimum Inhibitory Concentration (MIC) method on a VITEK 2 Compact 30 analyzer (bioMerieux, Marcy-l'Étoile, France). The antibiotics panel included the following drugs: amikacin, gentamicin, tobramycin, imipenem, meropenem, levofloxacin, ciprofloxacin, and trimethoprim/sulfomethoxazole. These antimicrobial compounds reflected those agents used for human therapy in the Russian Federation. We used the EUCAST clinical breakpoints, version 11.0 (https://www.eucast.org/clinical_breakpoints/, accessed on 20 December 2020), to interpret the results obtained.

4.2. DNA Isolation, Sequencing, and Genome Assembly

Nine samples were obtained from eight patients (5 males and 3 females) in various sources and hospital departments (Table 1) of a multidisciplinary federal medical center in Moscow, Russia, during the period 2017–2019. The age of the patients involved in this study ranged from 27 to 62 years with a median equal to 56.

The total number of isolates involved in the initial screening was 145, and 49 of them were sequenced. Earlier we have investigated the properties of the CRISPR/Cas arrays and systems for some of these isolates and a set of reference isolates from RefSeq [56]. Then we carefully selected nine isolates from the initial set, which represented the diversity of the *A. baumannii* international clones revealed in the hospital, and performed long-read sequencing for them. Our aim was not to capture the diversity of all strains found in the hospital during the study period, but rather to investigate the spread of international high-risk clones and to check the hypothesis of their endemicity for a particular region or country. Long-read sequencing allowed us to obtain the precise genome and plasmid structures, as well as to verify the locations of antibiotic resistance and virulence determinants, and to obtain complete cgMLST profiles for the selected representative isolates.

Genomic DNA was isolated with the DNeasy Blood and Tissue kit (Qiagen, Hilden, Germany). A Nextera™ DNA Sample Prep Kit (Illumina®, San Diego, CA, USA) was used for paired-end library preparation, and whole-genome sequencing (WGS) of the isolates on Illumina® Miseq and Hiseq platforms (Illumina®, San Diego, CA, USA).

Additional WGS was performed using the Oxford Nanopore MinION sequencing system (Oxford Nanopore Technologies, Oxford, UK). The same genomic DNA was used to prepare the MinION library with the Rapid Barcoding Sequencing kit SQK-RBK004 (Oxford Nanopore Technologies, Oxford, UK). The amount of initial DNA used for barcoding kit was 400 ng for each sample. The libraries were prepared according to the manufacturer's protocols, and were sequenced on R9 SpotON flow cell by initiating the standard 24 h sequencing protocol using the MinKNOW software (Oxford Nanopore Technologies, Oxford, UK).

Base calling of the raw MinION data was performed using Guppy Basecalling Software version 4.4.1 (Oxford Nanopore Technologies, Oxford, UK), and demultiplexing was made using Guppy barcoding software version 4.4.1 (Oxford Nanopore Technologies, Oxford, UK). Hybrid assemblies were obtained using short- and long-reads by Unicycler version 0.4.9-beta [57].

Genome assemblies were uploaded to NCBI Genbank under the following accession numbers: JAEPWJ000000000 (CriePir33), JAEPWI000000000 (CriePir87), JAEPWG000000000 (CriePir168), JAHHIS000000000 (CriePir254), JAEPWB000000000 (CriePir298), JAEPVY0000000 00 (CriePir306), JAEPVX000000000 (CriePir307), JAEPVW000000000 (CriePir308), and JAEPVV 000000000 (CriePir309).

4.3. Data Processing

The genomes assembled were processed using a custom software pipeline described earlier [33]. We used the Resfinder 4.0 database for antimicrobial gene identification (https://cge.cbs.dtu.dk/services/ResFinder/, accessed on 20 April 2021). VFDB [58] was used to search for the virulence factors http://www.mgc.ac.cn/VFs/main.htm (accessed on 20 April 2021).

Isolate typing was first performed by MLST using the Pasteur scheme. It was chosen for typing since, according to the Oxford scheme, five isolates possessed undetectable sequence types (STs) with a duplicated *gdhB* locus. Additional classification was made by using capsule synthesis loci (K-loci) [59] and lipooligosaccharide outer core loci (OCL) [60]. Detection of cgMLST profiles was performed using MentaList software (https://github.com/WGS-TB/MentaLiST, version 0.2.4, accessed on 10 June 2021), and the minimum spanning tree was build using PHYLOViz online (http://online.phyloviz.net, accessed on 10 June 2021).

CRISPRCasFinder [61] was used to identify the presence of CRISPR/Cas systems and spacers in the genomes studied. Anti-CRISPR elements were searched in AcrBank http://cefg.uestc.cn/anti-CRISPRdb (accessed on 20 April 2021).

5. Conclusions

In this study, we performed third-generation sequencing-based genomic epidemiology surveillance of clinical *A. baumannii* isolates from a multidisciplinary medical center in Moscow, Russia, obtained during 2017–2019. Surprisingly, we revealed that our isolates included 5 of the 9 commonly defined international clones of this important nosocomial pathogen. In addition, three isolates possessed singleton sequence types not clustered with the known lineages, including ST911, for which whole-genome data are not available yet. We presented a detailed analysis of the phenotypic antimicrobial resistance and genomic resistance determinants for all the isolates studies, as well as additional data for virulence factors, plasmids, and CRISPR arrays. We believe that these data will facilitate a better understanding of the clonal spreading and resistance acquisition of *A. baumannii* and further highlight the necessity of continuous genomic epidemiology surveillance for this problematic pathogen.

Supplementary Materials: The following are available online at https://www.mdpi.com/article/10.3390/antibiotics10081009/s1, Table S1. cgMLST profiles for clinical *A. baumannii* isolates included in the study, Table S2. Accession numbers for the virulence genes revealed in the isolates studied, Table S3. Plasmid description for the isolates studied, Table S4. Description and statistics for the CRISPR arrays revealed in clinical *A. baumannii* isolates included in the study, Figure S1. cgMLST tree for the isolates under study and their closest matches from RefSeq database.

Author Contributions: All authors contributed to this study; L.P. and M.Z. performed the clinical part of the experiment; Y.M. performed the sequencing part; A.S. analyzed the data and wrote the manuscript; V.A. supervised the project and obtained funding. All authors have read and agreed to the published version of the manuscript.

Funding: This work was supported by the Ministry of Science and Higher Education of the Russian Federation within the framework of a grant in the form of a subsidy for the creation and development of the «World-class Genomic Research Center for Ensuring Biological Safety and Technological Independence under the Federal Scientific and Technical Program for the Development of Genetic Technologies», agreement No. 075-15-2019-1666.

Institutional Review Board Statement: Ethical review and approval were waived for this study since the human samples were routinely collected and patients' data remained anonymous.

Informed Consent Statement: Patient consent was waived since the human samples were routinely collected, and patients' data remained anonymous. Patients codes reported in the manuscript are used for description purposes only and do not correspond to patient records within the hospital.

Data Availability Statement: The bacterial genomes presented in this study are openly available in NCBI Genbank under the following accession numbers: JAEPWJ000000000 (CriePir33), JAEPWI000000000 (CriePir87), JAEPWG000000000 (CriePir168), JAHHIS000000000 (CriePir254), JAEPWB000000000 (CriePir298), JAEPVY000000000 (CriePir306), JAEPVX000000000 (CriePir307), JAEPVW000000000 (CriePir308), and JAEPVV000000000 (CriePir309).

Conflicts of Interest: The authors declare no conflict of interest. The funders had no role in the design of the study; in the collection, analyses, or interpretation of data; in the writing of the manuscript, or in the decision to publish the results.

References

1. Armstrong, G.L.; MacCannell, D.R.; Taylor, J.; Carleton, H.A.; Neuhaus, E.B.; Bradbury, R.S.; Posey, J.E.; Gwinn, M. Pathogen Genomics in Public Health. *N. Engl. J. Med.* **2019**, *381*, 2569–2580. [CrossRef]
2. Gilchrist, C.A.; Turner, S.D.; Riley, M.F.; Petri, W.A., Jr.; Hewlett, E.L. Whole-genome sequencing in outbreak analysis. *Clin. Microbiol. Rev.* **2015**, *28*, 541–563. [CrossRef]
3. Maljkovic Berry, I.; Melendrez, M.C.; Bishop-Lilly, K.A.; Rutvisuttinunt, W.; Pollett, S.; Talundzic, E.; Morton, L.; Jarman, R.G. Next Generation Sequencing and Bioinformatics Methodologies for Infectious Disease Research and Public Health: Approaches, Applications, and Considerations for Development of Laboratory Capacity. *J. Infect. Dis.* **2020**, *221*, S292–S307. [CrossRef]
4. Hawken, S.E.; Snitkin, E.S. Genomic epidemiology of multidrug-resistant Gram-negative organisms. *Ann. N. Y. Acad. Sci.* **2019**, *1435*, 39–56. [CrossRef]
5. Palmieri, M.; D'Andrea, M.M.; Pelegrin, A.C.; Mirande, C.; Brkic, S.; Cirkovic, I.; Goossens, H.; Rossolini, G.M.; van Belkum, A. Genomic Epidemiology of Carbapenem- and Colistin-Resistant Klebsiella pneumoniae Isolates From Serbia: Predominance of ST101 Strains Carrying a Novel OXA-48 Plasmid. *Front. Microbiol.* **2020**, *11*, 294. [CrossRef] [PubMed]
6. Osei Sekyere, J.; Reta, M.A. Genomic and Resistance Epidemiology of Gram-Negative Bacteria in Africa: A Systematic Review and Phylogenomic Analyses from a One Health Perspective. *mSystems* **2020**, *5*, e00897-20. [CrossRef] [PubMed]
7. Talebi Bezmin Abadi, A.; Rizvanov, A.A.; Haertle, T.; Blatt, N.L. World Health Organization Report: Current Crisis of Antibiotic Resistance. *BioNanoScience* **2019**, *9*, 778–788. [CrossRef]
8. Fair, R.J.; Tor, Y. Antibiotics and bacterial resistance in the 21st century. *Perspect. Medicin. Chem.* **2014**, *6*, 25–64. [CrossRef] [PubMed]
9. Kuzmenkov, A.Y.; Trushin, I.V.; Vinogradova, A.G.; Avramenko, A.A.; Sukhorukova, M.V.; Malhotra-Kumar, S.; Dekhnich, A.V.; Edelstein, M.V.; Kozlov, R.S. AMRmap: An Interactive Web Platform for Analysis of Antimicrobial Resistance Surveillance Data in Russia. *Front. Microbiol.* **2021**, *12*, 620002. [CrossRef]
10. Magill, S.S.; Edwards, J.R.; Bamberg, W.; Beldavs, Z.G.; Dumyati, G.; Kainer, M.A.; Lynfield, R.; Maloney, M.; McAllister-Hollod, L.; Nadle, J.; et al. Multistate point-prevalence survey of health care-associated infections. *N. Engl. J. Med.* **2014**, *370*, 1198–1208. [CrossRef]
11. Lob, S.H.; Hoban, D.J.; Sahm, D.F.; Badal, R.E. Regional differences and trends in antimicrobial susceptibility of Acinetobacter baumannii. *Int. J. Antimicrob. Agents* **2016**, *47*, 317–323. [CrossRef]
12. Zhen, X.; Lundborg, C.S.; Sun, X.; Hu, X.; Dong, H. Economic burden of antibiotic resistance in ESKAPE organisms: A systematic review. *Antimicrob. Resist. Infect. Control.* **2019**, *8*, 137. [CrossRef]
13. Hamidian, M.; Nigro, S.J. Emergence, molecular mechanisms and global spread of carbapenem-resistant Acinetobacter baumannii. *Microb. Genom.* **2019**, *5*, e000306. [CrossRef]
14. Harding, C.M.; Hennon, S.W.; Feldman, M.F. Uncovering the mechanisms of Acinetobacter baumannii virulence. *Nat. Rev. Microbiol.* **2018**, *16*, 91–102. [CrossRef] [PubMed]

15. Yang, C.H.; Su, P.W.; Moi, S.H.; Chuang, L.Y. Biofilm Formation in Acinetobacter Baumannii: Genotype-Phenotype Correlation. *Molecules* **2019**, *24*, 1849. [CrossRef]
16. Holt, K.; Kenyon, J.J.; Hamidian, M.; Schultz, M.B.; Pickard, D.J.; Dougan, G.; Hall, R. Five decades of genome evolution in the globally distributed, extensively antibiotic-resistant Acinetobacter baumannii global clone 1. *Microb. Genom.* **2016**, *2*, e000052. [CrossRef]
17. Zarrilli, R.; Pournaras, S.; Giannouli, M.; Tsakris, A. Global evolution of multidrug-resistant Acinetobacter baumannii clonal lineages. *Int. J. Antimicrob. Agents* **2013**, *41*, 11–19. [CrossRef] [PubMed]
18. Al-Hassan, L.; Elbadawi, H.; Osman, E.; Ali, S.; Elhag, K.; Cantillon, D.; Wille, J.; Seifert, H.; Higgins, P.G. Molecular Epidemiology of Carbapenem-Resistant Acinetobacter baumannii From Khartoum State, Sudan. *Front. Microbiol.* **2021**, *12*, 628736. [CrossRef] [PubMed]
19. Shek, E.A.; Sukhorukova, M.V.; Edelstein, M.V.; Skleenova, E.Y.; Ivanchik, N.V.; Shajdullina, E.R.; Kuzmenkov, A.Y.; Dekhnich, A.V.; Kozlov, R.S.; Semyonova, N.V.; et al. Antimicrobial resistance, carbapenemase production, and genotypes of nosocomial Acinetobacter spp. isolates in Russia: Results of multicenter epidemiological study "MARATHON 2015–2016". *Clin. Microbiol. Antimicrob. Chemother.* **2019**, *21*, 171–180. [CrossRef]
20. Mayanskiy, N.; Chebotar, I.; Alyabieva, N.; Kryzhanovskaya, O.; Savinova, T.; Turenok, A.; Bocharova, Y.; Lazareva, A.; Polikarpova, S.; Karaseva, O. Emergence of the Uncommon Clone ST944/ST78 Carrying blaOXA-40-like and blaCTX-M-like Genes Among Carbapenem-Nonsusceptible Acinetobacter baumannii in Moscow, Russia. *Microb. Drug Resist.* **2017**, *23*, 864–870. [CrossRef] [PubMed]
21. Nodari, C.S.; Cayo, R.; Streling, A.P.; Lei, F.; Wille, J.; Almeida, M.S.; de Paula, A.I.; Pignatari, A.C.C.; Seifert, H.; Higgins, P.G.; et al. Genomic Analysis of Carbapenem-Resistant Acinetobacter baumannii Isolates Belonging to Major Endemic Clones in South America. *Front. Microbiol.* **2020**, *11*, 584603. [CrossRef]
22. Cerezales, M.; Xanthopoulou, K.; Wille, J.; Bustamante, Z.; Seifert, H.; Gallego, L.; Higgins, P.G. Acinetobacter baumannii analysis by core genome multi-locus sequence typing in two hospitals in Bolivia: Endemicity of international clone 7 isolates (CC25). *Int. J. Antimicrob. Agents* **2019**, *53*, 844–849. [CrossRef]
23. Higgins, P.G.; Hagen, R.M.; Kreikemeyer, B.; Warnke, P.; Podbielski, A.; Frickmann, H.; Loderstadt, U. Molecular Epidemiology of Carbapenem-Resistant Acinetobacter baumannii Isolates from Northern Africa and the Middle East. *Antibiotics* **2021**, *10*, 291. [CrossRef]
24. Shashkov, A.S.; Cahill, S.M.; Arbatsky, N.P.; Westacott, A.C.; Kasimova, A.A.; Shneider, M.M.; Popova, A.V.; Shagin, D.A.; Shelenkov, A.A.; Mikhailova, Y.V.; et al. Acinetobacter baumannii K116 capsular polysaccharide structure is a hybrid of the K14 and revised K37 structures. *Carbohydr. Res.* **2019**, *484*, 107774. [CrossRef]
25. Bonnin, R.A.; Potron, A.; Poirel, L.; Lecuyer, H.; Neri, R.; Nordmann, P. PER-7, an extended-spectrum beta-lactamase with increased activity toward broad-spectrum cephalosporins in Acinetobacter baumannii. *Antimicrob. Agents Chemother.* **2011**, *55*, 2424–2427. [CrossRef]
26. Naas, T.; Oueslati, S.; Bonnin, R.A.; Dabos, M.L.; Zavala, A.; Dortet, L.; Retailleau, P.; Iorga, B.I. Beta-lactamase database (BLDB)—structure and function. *J. Enzyme Inhib. Med. Chem.* **2017**, *32*, 917–919. [CrossRef]
27. Matthews, L.; Goodrich, J.S.; Weber, D.J.; Bergman, N.H.; Miller, M.B. The Brief Case: A Fatal Case of Necrotizing Fasciitis Due to Multidrug-Resistant Acinetobacter baumannii. *J. Clin. Microbiol.* **2019**, *57*, e01751-18. [CrossRef]
28. Hamidian, M.; Hawkey, J.; Wick, R.; Holt, K.E.; Hall, R.M. Evolution of a clade of Acinetobacter baumannii global clone 1, lineage 1 via acquisition of carbapenem- and aminoglycoside-resistance genes and dispersion of ISAba1. *Microb. Genom.* **2019**, *5*, e000242. [CrossRef]
29. Shelenkov, A.; Korotkov, A.; Korotkov, E. MMsat—a database of potential micro- and minisatellites. *Gene* **2008**, *409*, 53–60. [CrossRef]
30. Yeh, H.Y.; Awad, A. Genotyping of Campylobacter jejuni Isolates from Poultry by Clustered Regularly Interspaced Short Palindromic Repeats (CRISPR). *Curr. Microbiol.* **2020**, *77*, 1647–1652. [CrossRef]
31. Bartual, S.G.; Seifert, H.; Hippler, C.; Luzon, M.A.; Wisplinghoff, H.; Rodriguez-Valera, F. Development of a multilocus sequence typing scheme for characterization of clinical isolates of Acinetobacter baumannii. *J. Clin. Microbiol.* **2005**, *43*, 4382–4390. [CrossRef]
32. Diancourt, L.; Passet, V.; Nemec, A.; Dijkshoorn, L.; Brisse, S. The population structure of Acinetobacter baumannii: Expanding multiresistant clones from an ancestral susceptible genetic pool. *PLoS ONE* **2010**, *5*, e10034. [CrossRef] [PubMed]
33. Shelenkov, A.; Mikhaylova, Y.; Yanushevich, Y.; Samoilov, A.; Petrova, L.; Fomina, V.; Gusarov, V.; Zamyatin, M.; Shagin, D.; Akimkin, V. Molecular Typing, Characterization of Antimicrobial Resistance, Virulence Profiling and Analysis of Whole-Genome Sequence of Clinical Klebsiella pneumoniae Isolates. *Antibiotics* **2020**, *9*, 261. [CrossRef]
34. Higgins, P.G.; Prior, K.; Harmsen, D.; Seifert, H. Development and evaluation of a core genome multilocus typing scheme for whole-genome sequence-based typing of Acinetobacter baumannii. *PLoS ONE* **2017**, *12*, e0179228. [CrossRef] [PubMed]
35. Venditti, C.; Vulcano, A.; D'Arezzo, S.; Gruber, C.E.M.; Selleri, M.; Antonini, M.; Lanini, S.; Marani, A.; Puro, V.; Nisii, C.; et al. Epidemiological investigation of an Acinetobacter baumannii outbreak using core genome multilocus sequence typing. *J. Glob. Antimicrob. Resist.* **2019**, *17*, 245–249. [CrossRef]
36. Evans, B.A.; Amyes, S.G. OXA beta-lactamases. *Clin. Microbiol. Rev.* **2014**, *27*, 241–263. [CrossRef]

37. Gaiarsa, S.; Batisti Biffignandi, G.; Esposito, E.P.; Castelli, M.; Jolley, K.A.; Brisse, S.; Sassera, D.; Zarrilli, R. Comparative Analysis of the Two Acinetobacter baumannii Multilocus Sequence Typing (MLST) Schemes. *Front. Microbiol.* **2019**, *10*, 930. [CrossRef]

38. Cieslinski, J.M.; Arend, L.; Tuon, F.F.; Silva, E.P.; Ekermann, R.G.; Dalla-Costa, L.M.; Higgins, P.G.; Seifert, H.; Pilonetto, M. Molecular epidemiology characterization of OXA-23 carbapenemase-producing Acinetobacter baumannii isolated from 8 Brazilian hospitals using repetitive sequence-based PCR. *Diagn. Microbiol. Infect. Dis.* **2013**, *77*, 337–340. [CrossRef]

39. Wareth, G.; Brandt, C.; Sprague, L.D.; Neubauer, H.; Pletz, M.W. Spatio-Temporal Distribution of Acinetobacter baumannii in Germany-A Comprehensive Systematic Review of Studies on Resistance Development in Humans (2000–2018). *Microorganisms* **2020**, *8*, 375. [CrossRef]

40. Rafei, R.; Dabboussi, F.; Hamze, M.; Eveillard, M.; Lemarie, C.; Gaultier, M.P.; Mallat, H.; Moghnieh, R.; Husni-Samaha, R.; Joly-Guillou, M.L.; et al. Molecular analysis of Acinetobacter baumannii strains isolated in Lebanon using four different typing methods. *PLoS ONE* **2014**, *9*, e115969. [CrossRef]

41. Caldart, R.V.; Fonseca, E.L.; Freitas, F.; Rocha, L.; Vicente, A.C. Acinetobacter baumannii infections in Amazon Region driven by extensively drug resistant international clones, 2016–2018. *Mem. Inst. Oswaldo. Cruz.* **2019**, *114*, e190232. [CrossRef] [PubMed]

42. Jia, H.; Sun, Q.; Ruan, Z.; Xie, X. Characterization of a small plasmid carrying the carbapenem resistance gene bla OXA-72 from community-acquired Acinetobacter baumannii sequence type 880 in China. *Infect. Drug Resist.* **2019**, *12*, 1545–1553. [CrossRef] [PubMed]

43. Tomaras, A.P.; Dorsey, C.W.; Edelmann, R.E.; Actis, L.A. Attachment to and biofilm formation on abiotic surfaces by Acinetobacter baumannii: Involvement of a novel chaperone-usher pili assembly system. *Microbiology* **2003**, *149*, 3473–3484. [CrossRef] [PubMed]

44. Choi, A.H.; Slamti, L.; Avci, F.Y.; Pier, G.B.; Maira-Litran, T. The pgaABCD locus of Acinetobacter baumannii encodes the production of poly-beta-1-6-N-acetylglucosamine, which is critical for biofilm formation. *J. Bacteriol.* **2009**, *191*, 5953–5963. [CrossRef]

45. Sarshar, M.; Behzadi, P.; Scribano, D.; Palamara, A.T.; Ambrosi, C. Acinetobacter baumannii: An Ancient Commensal with Weapons of a Pathogen. *Pathogens* **2021**, *10*, 387. [CrossRef]

46. Liu, C.; Chang, Y.; Xu, Y.; Luo, Y.; Wu, L.; Mei, Z.; Li, S.; Wang, R.; Jia, X. Distribution of virulence-associated genes and antimicrobial susceptibility in clinical Acinetobacter baumannii isolates. *Oncotarget* **2018**, *9*, 21663–21673. [CrossRef]

47. He, X.; Lu, F.; Yuan, F.; Jiang, D.; Zhao, P.; Zhu, J.; Cheng, H.; Cao, J.; Lu, G. Biofilm Formation Caused by Clinical Acinetobacter baumannii Isolates Is Associated with Overexpression of the AdeFGH Efflux Pump. *Antimicrob. Agents Chemother.* **2015**, *59*, 4817–4825. [CrossRef]

48. Vazquez-Lopez, R.; Solano-Galvez, S.G.; Juarez Vignon-Whaley, J.J.; Abello Vaamonde, J.A.; Padro Alonzo, L.A.; Rivera Resendiz, A.; Muleiro Alvarez, M.; Vega Lopez, E.N.; Franyuti-Kelly, G.; Alvarez-Hernandez, D.A.; et al. Acinetobacter baumannii Resistance: A Real Challenge for Clinicians. *Antibiotics* **2020**, *9*, 205. [CrossRef]

49. Tang, J.; Chen, Y.; Wang, X.; Ding, Y.; Sun, X.; Ni, Z. Contribution of the AbaI/AbaR Quorum Sensing System to Resistance and Virulence of Acinetobacter baumannii Clinical Strains. *Infect. Drug Resist.* **2020**, *13*, 4273–4281. [CrossRef]

50. Schneidewind, L.; Kranz, J.; Tandogdu, Z. Rising significance of antibiotic stewardship in urology and urinary tract infections—a rapid review. *Curr. Opin. Urol.* **2021**, *31*, 285–290. [CrossRef]

51. Provenzani, A.; Hospodar, A.R.; Meyer, A.L.; Leonardi Vinci, D.; Hwang, E.Y.; Butrus, C.M.; Polidori, P. Multidrug-resistant gram-negative organisms: A review of recently approved antibiotics and novel pipeline agents. *Int. J. Clin. Pharm.* **2020**, *42*, 1016–1025. [CrossRef]

52. Shelenkov, A.; Slavokhotova, A.; Odintsova, T. Predicting Antimicrobial and Other Cysteine-Rich Peptides in 1267 Plant Transcriptomes. *Antibiotics* **2020**, *9*, 60. [CrossRef]

53. Ghosh, C.; Sarkar, P.; Issa, R.; Haldar, J. Alternatives to Conventional Antibiotics in the Era of Antimicrobial Resistance. *Trends Microbiol.* **2019**, *27*, 323–338. [CrossRef]

54. Bebbington, C.; Yarranton, G. Antibodies for the treatment of bacterial infections: Current experience and future prospects. *Curr. Opin. Biotechnol.* **2008**, *19*, 613–619. [CrossRef]

55. Uppu, D.S.; Manjunath, G.B.; Yarlagadda, V.; Kaviyil, J.E.; Ravikumar, R.; Paramanandham, K.; Shome, B.R.; Haldar, J. Membrane-active macromolecules resensitize NDM-1 gram-negative clinical isolates to tetracycline antibiotics. *PLoS ONE* **2015**, *10*, e0119422. [CrossRef]

56. Tyumentseva, M.; Mikhaylova, Y.; Prelovskaya, A.; Tyumentsev, A.; Petrova, L.; Fomina, V.; Zamyatin, M.; Shelenkov, A.; Akimkin, V. Genomic and Phenotypic Analysis of Multidrug-Resistant Acinetobacter baumannii Clinical Isolates Carrying Different Types of CRISPR/Cas Systems. *Pathogens* **2021**, *10*, 205. [CrossRef] [PubMed]

57. Wick, R.R.; Judd, L.M.; Gorrie, C.L.; Holt, K.E. Unicycler: Resolving bacterial genome assemblies from short and long sequencing reads. *PLoS Comput. Biol.* **2017**, *13*, e1005595. [CrossRef] [PubMed]

58. Liu, B.; Zheng, D.; Jin, Q.; Chen, L.; Yang, J. VFDB 2019: A comparative pathogenomic platform with an interactive web interface. *Nucleic Acids Res.* **2019**, *47*, D687–D692. [CrossRef] [PubMed]

59. Arbatsky, N.P.; Shneider, M.M.; Dmitrenok, A.S.; Popova, A.V.; Shagin, D.A.; Shelenkov, A.A.; Mikhailova, Y.V.; Edelstein, M.V.; Knirel, Y.A. Structure and gene cluster of the K125 capsular polysaccharide from Acinetobacter baumannii MAR13-1452. *Int. J. Biol. Macromol.* **2018**, *117*, 1195–1199. [CrossRef]

60. Wyres, K.L.; Cahill, S.M.; Holt, K.E.; Hall, R.M.; Kenyon, J.J. Identification of Acinetobacter baumannii loci for capsular polysaccharide (KL) and lipooligosaccharide outer core (OCL) synthesis in genome assemblies using curated reference databases compatible with Kaptive. *Microb. Genom.* **2020**, *6*, e000339. [CrossRef]
61. Couvin, D.; Bernheim, A.; Toffano-Nioche, C.; Touchon, M.; Michalik, J.; Neron, B.; Rocha, E.P.C.; Vergnaud, G.; Gautheret, D.; Pourcel, C. CRISPRCasFinder, an update of CRISRFinder, includes a portable version, enhanced performance and integrates search for Cas proteins. *Nucleic Acids Res.* **2018**, *46*, W246–W251. [CrossRef] [PubMed]

antibiotics

Article

Molecular Detection, Serotyping, and Antibiotic Resistance of Shiga Toxigenic *Escherichia coli* Isolated from She-Camels and In-Contact Humans in Egypt

Mohamed Said Diab [1], Reda Tarabees [2], Yasser F. Elnaker [3], Ghada A. Hadad [4], Marwa A. Saad [5], Salah A. Galbat [3], Sarah Albogami [6], Aziza M. Hassan [6], Mahmoud A. O. Dawood [7,8,*] and Sabah Ibrahim Shaaban [9]

[1] Department of Animal Hygiene and Zoonoses, Faculty of Veterinary Medicine, New Valley University, El Kharga 72511, Egypt; mohameddiab333@gmail.com
[2] Department of Bacteriology, Mycology, and Immunology, Faculty of Veterinary Medicine, University of Sadat City, Sadat City 32897, Egypt; reda.tarabees@vet.usc.edu.eg
[3] Department of Animal Medicine (Infectious Diseases), Faculty of Veterinary Medicine, The New Valley University, El Kharga 72511, Egypt; yasserelnaker@yahoo.com (Y.F.E.); salahgalbat@yahoo.com (S.A.G.)
[4] Department of Animal Hygiene and Zoonoses, Faculty of Veterinary Medicine, University of Sadat City, Sadat City 32897, Egypt; ghadavet@yahoo.com
[5] Department of Food Control, Faculty of Veterinary Medicine, Menofia University, Menofia, ShebinAl-Kom 32511, Egypt; Drmarwa2200@gmail.com
[6] Department of Biotechnology, College of Science, Taif University, P.O. Box 11099, Taif 21944, Saudi Arabia; dr.sarah@tu.edu.sa (S.A.); a.hasn@tu.edu.sa (A.M.H.)
[7] Department of Animal Production, Faculty of Agriculture, Kafrelsheikh University, Kafrelsheikh 33512, Egypt
[8] The Center for Applied Research on the Environment and Sustainability, The American University in Cairo, Cairo 11835, Egypt
[9] Department of Animal Hygiene and Zoonoses, Faculty of Veterinary Medicine, Damanhour University, Damanhour 22511, Egypt; sabah_ibrahim@vetmed.dmu.edu.eg or sabahibrahim2010@gmail.com
* Correspondence: Mahmouddawood55@gmail.com or mahmoud.dawood@agr.kfs.edu.eg

Citation: Diab, M.S.; Tarabees, R.; Elnaker, Y.F.; Hadad, G.A.; Saad, M.A.; Galbat, S.A.; Albogami, S.; Hassan, A.M.; Dawood, M.A.O.; Shaaban, S.I. Molecular Detection, Serotyping, and Antibiotic Resistance of Shiga Toxigenic *Escherichia coli* Isolated from She-Camels and In-Contact Humans in Egypt. *Antibiotics* **2021**, *10*, 1021. https://doi.org/10.3390/antibiotics10081021

Academic Editors: Manuela Oliveira and Elisabete Silva

Received: 3 July 2021
Accepted: 18 August 2021
Published: 23 August 2021

Publisher's Note: MDPI stays neutral with regard to jurisdictional claims in published maps and institutional affiliations.

Abstract: This study aims to determine the prevalence of STEC in she-camels suffering from mastitis in semi-arid regions by using traditional culture methods and then confirming it with Serological and molecular techniques in milk samples, camel feces, as well as human stool samples for human contacts. In addition, an antibiotic susceptibility profile for these isolates was investigation. Mastitic milk samples were taken after California Mastitis Test (CMT) procedure, and fecal samples were taken from she-camels and human stool samples, then cultured using traditional methods to isolate *Escherichia coli*. These isolates were initially classified serologically, then an mPCR (Multiplex PCR) was used to determine virulence genes. Finally, both camel and human isolates were tested for antibiotic susceptibility. Out of a total of 180 she-camels, 34 (18.9%) were mastitic (8.3% clinical and 10.6% sub-clinical mastitis), where it was higher in camels bred with other animals. The total presence of *E. coli* was 21.9, 13.9, and 33.7% in milk, camel feces, and human stool, respectively, whereas the occurrence of STEC from the total *E. coli* isolates were 36, 16, and 31.4% for milk, camel feces, and stool, respectively. Among the camel isolates, stx_1 was the most frequently detected virulence gene, while *hlyA* was not detected. The most detected virulence gene in human isolates was stx_2 (45.5%), followed by stx_1. Camel STEC showed resistance to Oxytetracycline only, while human STEC showed multiple drug resistance to Amoxicillin, Gentamycin, and Clindamycin with 81.8, 72.7, and 63.6%, respectively. Breeding camels in semi-arid areas separately from other animals may reduce the risk of infection with some bacteria, including *E. coli*; in contrast, mixed breeding with other animals contributes a significant risk factor for STEC emergence in camels.

Keywords: antibiotic resistance; camel; *Escherichia coli*; domestic; milk; mastitis

1. Introduction

Camels are characterized by their remarkable ability to adapt to the extreme desert ecosystem and their high resistance to many pathogenic microorganisms (MOs) compared to other domesticated animals in the same area [1]. Dromedary camels contribute strongly to human survival in the Middle East, and North and East Africa regions. The main reason for raising camels is to produce milk, as camels produce more milk for a longer period when compared to other dairy animals. However, its role in human transportation and as an essential source of meat cannot be ignored [2–4]. Camel milk is one of the main and important components in the human diet in these regions because it has a high nutritious value such as a high proportion of vitamin C, antibacterial substance, lactoferrin as well as some minerals, and minimum sugar and cholesterol content in comparison to cow milk [5,6].

Few published scientific studies are dealing with the causative agents of camel diseases, including mastitis [7], which is one of the most important diseases that affect dairy animals, that results in severe economic losses, including a decrease in milk yield, and the cost of treatment in addition to the public health risks [8–10]. Mastitis is an uncommon disease in camels compared to cows, but its incidence often increases with several things, including teat deformities, hand milking, and herd management [6]. On the other hand, Bessalah et al. [11] pointed to camel diarrhea as the main cause of economic loss associated with poor growth, medication costs, and animal death. Mastitis has extreme zoonotic and economic importance since it causes multiple hazardous effects on human health and animal production. Moreover, in these regions, the daily consumption of camel milk mainly occurs in the raw form [12,13].

Shiga Toxigenic *Escherichia coli* (STEC) is a significant foodborne zoonotic pathogen responsible for mild to severe diarrhea, hemorrhagic colitis, and hemolytic uremic syndrome. STEC virulence factors are derived from Shiga toxin genes (stx_1 and stx_2), which are the chief factors accountable for the clinical signs, intimin (*eae*), and hemolysin (*hlyA*) [14,15]. Hand-to-mouth transfer, considered as direct contact with farm animals, is the dominant mode of STEC transmission to human. Ruminants, mainly cattle, are considered the primary source of STEC infection for humans [16,17]; some authors exclude the role of camels [18,19]. Diagnostic methods using molecular techniques are faster and more accurate than traditional culturing methods for determining the different bacterial species [20].

The excessive and misuse of antibiotics in humans, animals, and plants was considered one of the main contributors to increasing the incidence of multidrug-resistant bacteria [21–23]. In the past, some authors asserted that there were no multiple drug-resistant bacteria among the causes of mastitis [24,25]. However, recently, bacteria such as *E. coli* have been discovered that are resistant to many antibiotics, which may be transmitted from milk-producing cows to humans [26,27]. Little information is presented about antimicrobial resistance among pathogenic MOs in camel [28]. Even if the resistance rates to antibiotics are relatively low, it can be dangerous due to the possibility of the transmission and spread of resistance genes between strains [29]. The nomadic nature of this region and the reliance on medicinal plants as natural antibacterial agents may have been an influential factor in the discovery of low levels of multi-antibiotic resistance [6,30–32]. This may have a positive impact on both veterinary and public health [33]. Consequently, this work aimed to study the role of *STEC* in mastitis and diarrhea in she-camels and its incidence among human beings in the same area. Additionally, antibiotic susceptibility tests for the isolates were performed.

2. Results

The data presented in Table 1 revealed that the general occurrence of camel mastitis was 18.9% (8.3% clinical and 10.6% subclinical mastitis). Moreover, the results presented in Table 2 showed that the prevalence of clinical and subclinical mastitis among she-camels reared separately with no contact with other animal species was 4.1 and 5.6%, respectively, while the rates of infection in camels raised with other animal species increased, reaching

11.1 and 13.9%, respectively. There was a statistically significant difference between the two breeding systems, either in separate or mixed breeding.

Table 1. Occurrence of clinical and subclinical mastitis in she-camel.

Milk	180 she-camels	Total (34/180) = 18.9%
		Clinical mastitis 15 (8.3%)
		Subclinical 19 (10.6%)
	720 milk samples per quarter level (180 animals * 4 quarters)	Clinical mastitis 43 (5.9%)
		subclinical 71 (9.9%)
Fecal Samples	180	Diarrhea 9 (5%)
		Normal 171 (95%)

Table 2. Occurrence of *E. coli* in mastitic she-camel's milk in relation to camel breeding system (mixed with other species).

Types of Mastitis	Separate (no 72)		Mixed Breeding (no 108)		Chi-Square Value	*p*-Value
	No.	%	No.	%		
Clinical mastitis	3	4.1	12	11.1	6.58	0.04
Subclinical mastitis	4	5.6	15	13.9		
Total	7	9.7	27	25		

p-value is significant at <0.05.

The results presented in Table 3 showed that *E. coli* was isolated using conventional culture methods from 25.6 and 19.7% of the examined clinically and sub-clinically mastitic camel's milk. Concerning the isolation of *E. coli* from the fecal samples, our results showed that *E. coli* was isolated from 44.4 and 12.3% of the examined fecal samples collected from diarrheic camels and apparently healthy she-camels, correspondingly, as shown in Table 3.

Table 3. Isolation of *E. coli* from milk samples/quarter and fecal samples of the examined she-camels.

Camel Samples	*E. coli* Conventional Isolation		STEC (PCR)/Total Cases		STEC (PCR)/*E. coli* Isolates	
	No.	%	No.	%	No.	%
Clinical mastitis/quarter n = 43	11	25.6	3	6.9	3	27.3
Subclinical/quarter n= 71	14	19.7	6	8.5	6	42.9
Total/quarter n = 114	25	21.9	9	7.9	9	36
Diarrhea n = 9	4	44.4	1	11.1	1	25
Normal feces n = 171	21	12.3	3	1.8	3	14.3
Total n = 180	25	13.9	4	2.2	4	16

Regarding human stool samples (Table 4), our results showed that *E. coli* was isolated from 23.2 and 37.2% of the examined stool samples collected from contact and non-contact individuals, respectively, with no statistically significant difference ($p > 0.05$). Among these isolates, STEC represented 16.7 and 34.5%, respectively. Concerning the seasonal prevalence of STEC, our results presented in Table 5 revealed a higher prevalence of STEC in a cold climate than in hot climates.

Table 4. Isolation of *E. coli* from human stool samples.

Human Samples	*E. coli* Conventional Isolation		STEC (PCR)/Total Cases		STEC (PCR)/*E. coli* Isolates	
	No.	%	No.	%	No.	%
Contact $n = 26$	6	23.2	1	3.8	1	16.7
Non-contact $n = 78$	29	37.2	10	12.8	10	34.5
Total $n = 104$	35	33.7	11	10.6	11	31.4
Chi-square value		1.73		1.66		1.66
p-value		0.19		0.18		0.18

Table 5. Occurrence of STEC in relation to the hot and cold season.

	Hot Climate	Cold Climate
Clinical mastitis	0	3
subclinical	1	5
Diarrhea	0	1
Normal feces	0	0
Human isolates	1	10

Table 6 showed that six different STEC serotypes were recovered from camel samples, including O26, O45, O103, O111, O121, and O145 in percentages of 6, 2, 2, 6, 4, and 6%, respectively, whereas O26, O45, O103, and O145 serotypes were recovered from human stool samples in percentages of 11.4, 8.6, 5.7, and 5.7%, respectively.

Table 6. Serotyping of *E. coli* isolates.

Species	Serotypes of STEC					
	O26 (%)	O45 (%)	O103 (%)	O121 (%)	O145 (%)	O111 (%)
Camel isolates $n= 13$	3 (6)	1 (2)	1 (2)	2 (4)	3 (6)	3 (6)
Human isolates $n = 11$	4 (11.4)	3 (8.6)	2 (5.7)	0	2 (5.7)	0
Total $n = 24$	7	4	3	2	5	3

Table 7 showed that 13 (26%) and 11 (31.4%) of the examined *E. coli* isolates recovered from camel and human samples, respectively, were positive for at least one of the examined genes for *STEC*. Among the tested camel isolates, the most prevalent virulence factors were stx_1, *Eae*, and stx_2 by rates of 46.2, 30.7, and 23.1%, respectively.

Table 7. Occurrence of virulence factors in relation to isolates.

Species		Virulence Genes							
		stx_1	stx_2	stx_1 & stx_2	*eae*	*hlyA*	stx_1 + *eae*	stx_2 + *eae*	stx_1 & stx_2 + *eae*
STEC Camel isolates (13)	No.	6	3	4	4	0	3	1	0
	%	46.2	23.1	30.7	30.7	0	23.1	7.7	0
STEC human isolates (11)	No.	4	5	1	5	3	1	2	1
	%	36	45.5	9	45.5	27.3	9	18.1	9

In the present study, 13 and 11 of the STEC isolates recovered from camel and human samples, respectively, were screened for their antimicrobial susceptibility, as shown in Table 8. The current study results showed that the camel STEC isolates were sensitive to the most tested antibiotics, except for Oxytetracycline, to which the isolates showed resistance with 53.8%, while the human isolates of STEC showed the highest resistance to Amoxicillin, Gentamycin, and Clindamycin with ratios of 81.8, 72.7, and 63.6, respectively.

Table 8. Antibiotic sensitivity test for *STEC* isolates against different antibiotics using CLSI breakpoint [34].

Antibacterial Agent	No. of Resistance among Camel Isolates				No. of Resistance among Human Isolates			
	No.	R (%)	I (%)	S (%)	No.	R (%)	I (%)	S (%)
Streptomycin		1(7.7)	2(15.4)	10(77)		5 (45.5)	5(45.5)	1(9)
Gentamycin		1(7.7)	4(30.8)	8(61.5)		8 (72.7)	1(9)	2(18.2)
Clindamycin		0(0)	1(7.7)	12(92.3)		7 (63.6)	2(18.2)	2(18.2)
Amoxicillin	13	0(0)	2(15.4)	11(84.6)	11	9 (81.8)	1(9)	1(9)
Ampicillin		1(7.7)	3(23.1)	9(69.2)		4 (36.4)	1(9)	6(54.5)
Oxytetracycline		7(53.8)	4(30.8)	2(15.4)		2 (18.2)	4(36.4)	5(45.5)
Ciprofloxacin		1(7.7)	3(23.1)	9(69.2)		7 (63.6)	1(9)	3(27.3)

R = resistant; I = intermediate; S = susceptible.

3. Discussion

There is a dearth of information on STEC epidemiology in humans, food, and animals in Sub-Saharan Africa, and the current knowledge of STEC sources needs to be further improved [17]. Similarly, there is limited information on the occurrence and the characteristics of *STEC* in African camels. Therefore, the current study was undertaken to estimate STEC incidence in the mastitic milk and fecal samples of dromedary camels and in-contact human stool. In addition, the isolates were further characterized for the presence of some virulence encoding genes and antibiogram sensitivity patterns.

In the present study, a total of 180 she-camels were investigated for the presence of clinical and sub-clinical mastitis; the results revealed that the general occurrence of camel mastitis was 18.9% (8.3% clinical and 10.6% subclinical mastitis). However, the general occurrence on the udder quarters level was 6.9%. These results were nearly similar to Jilo et al. [6], who stated that subclinical mastitis was more prevalent than clinical mastitis. Higher results, 26.3%, were reported by Balemi et al. [35]. Moreover, the results showed that clinical and subclinical mastitis prevalence among she-camels reared separately with no contact with other animal species was 4.1 and 5.6%, respectively, while the infection rates in camels reared with other animals increased, reaching 11.1 and 13.9%, respectively. There was a statistically significant difference between the two breeding systems, either in separate or mixed breeding. Similar findings were validated by Clement et al. [23] due to the possibility of *STEC* cross-transmission between cattle and camels. Furthermore, the hygienic conditions of the camels' housing and milking conditions were pursued by the owners.

E. coli is a Gram-negative rod, representing an important component of the microbiota of mammals and birds. However, several strains of *E. coli*, mainly diarrheagenic *E. coli*, are pathogenic to human and animals and cause several gastrointestinal disorders, including diarrhea [36]. Despite the seriousness of diarrheagenic *E. coli*, especially *STEC*, the studies conducted in Egypt were limited to cattle and sheep [37–39], compared with those conducted on camels. Therefore, the milk samples collected from clinically and sub-clinically mastitic she-camels and feces were further examined for the presence of *E. coli*. The results showed that *E. coli* was isolated using conventional culture methods from 25.6 and 19.7% of the examined clinically and sub-clinically mastitic camel's milk. These findings are lower than those previously obtained by Abo Hashem et al. [40], who reported that *E. coli* was the most predominant isolated bacteria from she-camel's milk with isolation rates of 35.4 and 27% from apparently healthy and mastitic she-camel's milk, respectively. Concerning the isolation of *E. coli* from the fecal samples, our results showed that *E. coli* was isolated from 44.4 and 12.3% of the examined fecal samples collected from diarrheic camels and apparently healthy she-camels, correspondingly, as shown in Table 3. Similar detection rates of *E. coli* from she-camels were observed by Al Humam [41], who detected isolates in 26% of cases. Contrariwise, these findings were lower when compared with those formerly reported by El-Hewairy et al. [42] and Al-Ajmi et al. [43]. Conversely, our findings are higher than those reported by Shahein et al. [7], where *E. coli* was isolated from 17.1% of the examined fecal samples collected from diarrheic camels. Several studies were undertaken to assess the prevalence of *E. coli* in fecal samples collected from diarrheic camels in Qatar [44], United Arab Emirates [43], Kenya [45], and Nigeria at 3.8% [46]. However, El-Sayed et al. [18] failed to detect any *STEC* from camel feces. These differences in findings could be attributed to the area of samples collections and the hygienic conditions of the housing and the milking procedures.

Regarding the human stool samples (Table 4), our results showed that *E. coli* was isolated from 23.2 and 37.2% of the examined stool samples collected from contact and non-contact individuals, respectively. Among these isolates, STEC represented 16.7 and 34.5%, respectively, with no statistically significant difference. These findings are similar to those reported by EL-Alfy et al. [47], where *E. coli* was isolated from 31.4% of the examined diarrheic human stools. On the contrary, Ramadan et al. [48] stated that *E. coli* was isolated from 58.6 and 71.4% of the examined diarrheic and healthy individuals' fecal samples, respectively.

Concerning the seasonal prevalence of STEC, our results revealed a higher prevalence of *STEC* in a cold climate than in hot climates. A similar result was reported by Persson et al. [49], who declared that the prevalence of STEC was more prevalent in the wet season. These findings are inconsistent with those of Monaghan et al. [50] and Moses et al. [51], who reported an increased prevalence of STEC in the summer–early autumn among the camel population. These differences could be attributed to the management process and the isolation techniques used in different laboratories. However, further investigations are required to declare the effect of seasons on the prevalence of STEC.

Our results showed that six different STEC serotypes were recovered from camel samples, including O26, O45, O103, O111, O121, and O145 in percentages of 6, 2, 2, 6, 4 and 6%, respectively, whereas the O26, O45, O103, and O145 serotypes were recovered from human stool samples in percentages of 11.4, 8.6, 5.7, and 5.7%, respectively. These results are nearly comparable to those obtained by Shahein et al. [7], who isolated several *E. coli* serotypes from fecal samples of camel neonates, including O26, O103, O111, and O45 in a percentage of 33.3, 25, 25, and 16.7%, respectively, and Bakhtiari et al. [52], who concluded that the most recovered *STEC* serotypes from human isolates were O26, O45, O103, O111, O121, and O145.

STEC represents a significant health problem worldwide as it is accountable for an estimated 2,801,000 acute illnesses yearly [17]. STEC causes many infections in humans, including gastrointestinal illnesses including non-bloody or bloody diarrhea, hemorrhagic colitis, and hemolytic uremic syndrome [53], which has been infrequently identified in camels. The transmission of STEC usually occurs through contaminated foods, water, and person-to-person spread [54,55].

Our results showed that 13 (26%) and 11 (31.4%) of the examined *E. coli* isolates recovered from camel and human samples, respectively, were positive for at least one of the examined genes for STEC. Among the tested camel isolates, the most prevalent virulence factors were stx_1, *eae*, and stx_2 by rates of 46.2, 30.7, and 23.1%, respectively. These results were similar to the finding of Ranjbar et al. [56] and Rashid et al. [57], who stated that stx_1 was the most common virulence gene of STEC, and Mashak [58], who stated that the presence of this large combination of virulence factors increases the pathogenicity of the isolates. In contrast, none of the tested camel isolates were found to have the *hlyA* encoding gene. Despite this, Adamu et al. [46] found that virulence genes were present in substantial amounts in camel STEC and that stx_1 and stx_2 were present in 43.5% of the tested isolates. In addition, the *hlyA* gene was present in 69.6% of the isolates. On the other side, stx_2 was shown to be the most frequently detected in human isolates (45.5%), which is consistent with the findings of Miara et al. [31], Hakim et al. [39], and Adamu et al. [46].

The extensive use of antibiotics in treating infectious diseases and as feed additives has resulted in the emergence of multi-drug-resistant bacteria [59–62]. The emergence of multi-drug-resistant STEC is one of the concerns of health and food safety authorities worldwide [63,64], as the resistance genes can be reproduced and transmitted not only to other bacteria but also to other hosts, including humans. Previous reports showed that antibiograms are considered more reliable for detecting antibiotic resistance than genotypic resistance gene detection [65]. In the present study, 13 and 11 *STEC* isolates were recovered from camel and human samples, which were screened for antimicrobial susceptibility. The current study results showed that the camel *STEC* isolates were sensitive to the most tested antibiotics, except Oxytetracycline, to which the isolates showed resistance with 53.8%, while the human isolates of STEC showed the highest resistance to Amoxicillin, Gentamycin, and Clindamycin, with ratios of 81.8, 72.7, and 63.6, respectively. The closest results to this study for the resistant strains of several human isolates were reported by Momtaz et al. [66], who determined that the isolates were more resistant to Oxytetracycline 86%, and Ranjbar et al. [67], Gentamycin, Ciprofloxacin, and aminoglycosides. Higher results for antibiotic resistance were observed by Ababu et al. [68], who noted that the resistance of the isolates for both Oxytetracycline and Gentamycin was 100%. On the other hand, Al-Ajmi et al. [43] stated that 100% of STEC isolates were susceptible to Ciprofloxacin and 84% for Amoxicillin. The relatively little discovery of multiple drug-resistant human isolates to many antibiotics in this study may be due to the dependence of people in this region on traditional methods, which may have a prominent effect on maintaining human health [32,33].

Small proportions of the resistance of camel isolates to many antibiotics may be due to the nature of their breeding in these semi-arid desert areas and the lack of excessive use of antibiotics, whether in treatment or as growth stimulants, except for Oxytetracycline [58,66]. A high resistance to Oxytetracycline among camel isolates in our study was discovered, possibly because of the extensive use of these broad-spectrum antibiotics by paramedical personnel and camel holders. None of the isolates were resistant to Clindamycin, which is not surprising because there is no trade medicine for veterinary use that contains Clindamycin for large animal treatment as an active ingredient in Egypt.

4. Materials and Methods

4.1. Study Area and Animals

This study was conducted in Wadi El-Natroun, which is a semi-arid area in El-Behira governorate, Egypt, located in the Western desert, which is located 23 m (75 ft) below sea level and 38 m (125 ft) below the Nile River level. This study was conducted on 180 she-camels and humans in contact with these animals in the same area.

4.2. Sampling

Milk samples: Between 2020 and 2021, 720 milk samples were collected from 180 she-camels (4 udder quarters per animal). The camels were randomly selected, as they are bred sporadically in this semi-arid nomadic region and feed mainly on the grasses that grow in it. A part of each milk sample was tested using the California mastitis test (CMT), and the other portion of milk samples were placed directly in the icebox and sent to the laboratory with minimum delay. The CMT test is used to determine whether or not mastitis is present. Differentiation between subclinical and clinical mastitis was based on the apparent symptoms of mastitis (e.g., swelling and redness of the udder in addition to milk clotting).

Fecal samples: A total of 180 fecal samples were collected from the examined she-camels using rectal swabs in order to reduce potential environmental contamination, then placed directly in an icebox and sent to the laboratory as soon as possible.

Stool samples: A total of 104 stool samples were collected from people living in the same breeding areas as these camels. All the people in this work live in the same study area, some of them are in direct contact with the tested camels, and they carry out various care operations such as milking, providing them with food, and cleaning, and their number is 26. Others live in the same breeding areas only, but they are not in direct contact with the camels, and their number is 78. Then, the samples were placed directly in an icebox and sent to the laboratory with minimum delay. CMT was performed according to the procedure of Hoque et al. [69].

4.3. Isolation and Identification of E. coli

Of the tested milk samples, 25 mL was added to 225 mL of buffered peptone water. Additionally, the fecal and stool swabs were immersed in Macconkey broth. The samples were incubated aerobically at 37 °C/24 h. After that, the samples were streaked aerobically on Macconkey agar media at 37 °C/24 h; the suspected colonies were picked up and re-streaked on EMB at 37 °C/24 h for further processing purification. The presumptive green sheen metallic colonies were biochemically tested according to Quinn et al. [70]. The suspected lactose fermenter colonies were first picked up from Macconkey agar media then re-cultured on EMB for further purification. Finally, a pure separate colony was picked up for further investigation and identification.

4.4. Serotyping

Serological identification of E. coli isolates was performed according to Kok et al. [71].

4.5. Procedures for Determination of O-Antigen Group

Two separate drops of saline were put on a glass slide, and a portion of the colony from the suspected culture was emulsified with the saline solution to give a smooth, reasonably dense suspension. To one suspension, control, one loopful of saline was added and mixed. One loopful of the undiluted antiserum was added to the other suspension and tilted back and forward for one minute. Agglutination was observed using indirect lighting over a dark background. When a colony gave a strongly positive agglutination with one of the pools of polyvalent serum, a different portion of it was inoculated onto a nutrient agar slant and incubated at 37°C for 24 h to grow as a culture for testing with monovalent sera. A heavy suspension of bacteria from each slope culture was prepared in saline, and slide agglutination tests were performed with the diagnostic sera to identify the O-antigen.

4.6. PCR Template Preparation

One or two colonies of each confirmed *STEC* isolate were thoroughly mixed in 1 mL of distilled water then boiled for 10 min. The boiled suspension was centrifuged at 1200 rpm/3 min, then 1 μL of supernatant was used as a DNA template.

PCR procedure was carried out in a total volume of 20 μL. Each 20-milliliter PCR reaction mixture contained 10 mL of the 2X Fast Cycling PCR master mix (Qiagen Fast Cycling PCR Kit, Qiagen, Valencia, CA, USA); 4 mL of the primer master mix (stx_1, stx_2, *eae*, and *hlyA*) (Table 9); 5 mL of DNase, RNase-free water; and 1 ml of template DNA (200 e900 ng/mL). The reaction was performed in an Applied Biosystems 2720 thermal cycler under the following conditions.

Table 9. Primers used were supplied from Metabion (Germany).

Target Gene	Primer Sequence (5′–3′)	Fragment Size (bp)	
stx_1	F: ATAAATCGCCATTCGTTGACTAC R: AGAACGCCCACTGAGATCATC	180	
stx_2	F: GGCACTGTCTGAAACTGCTCC R: TCGCCAGTTATCTGACATTCTG	255	[72]
eae	F: GACCCGGCACAAGCATAAGC R: CCACCTGCAGCAACAAGAGG	384	
hlyA	F: GCATCACAAGCGTACGTTCC R: AATGAGCCAAGCTGGTTAAGCT	534	

Samples were subjected to 35 PCR cycles, each consisting of 1 min of denaturation at 95 °C; 2 min of annealing at 65 °C for the first 10 cycles, decrementing to 60 °C by cycle 15; and 1.5 min of elongation at 72 °C, incrementing to 2.5 min from cycles 25 to 35. Amplified DNA fragments were resolved using gel electrophoresis.

4.7. Antibiotic Susceptibility

STEC isolates were tested against the different antibiotics according to the CLSI breakpoint [34]. We limited the design of the experiment to the group of antibiotics used in the place of the study, knowing that the Bedouin nature makes them more inclined to use medicinal herbs in treatment. The tested antibiotics were Streptomycin (10 μg/disk), Gentamycin (10 μg/disk), Clindamycin (2 μg/disk), Amoxicillin (30 μg), Ampicillin (10 μg/disk), Oxytetracycline (30 μg), and Ciprofloxacin (5 μg/disk) (Table 10).

Table 10. Interpretation criteria.

Antimicrobial Agent	Disk Content	Zone Diameter Interpretive Criteria (Nearest Whole mm)		
		S	I	R
Clindamycin	2 μg			
Ampicillin	10 μg	$\geq$19	16–18	$\leq$15
Gentamycin	10 μg	$\geq$17	14–16	$\leq$13
Streptomycin	10 μg	$\geq$15	13–14	$\leq$12
Tetracycline	30 μg	$\geq$15	12–14	$\leq$11
Ciprofloxacin	5 μg	$\geq$15	12–14	$\leq$11
Amoxycillin	30 μg	$\geq$31	21–30	$\leq$20

R = resistant; I = intermediate; S = susceptible.

4.8. Statistical Analysis

The chi-square test was employed to compare differences between different values. A p-value of <0.05 was considered to indicate statistically significant differences.

Author Contributions: Conceptualization, M.S.D., R.T., Y.F.E., G.A.H., M.A.S., S.A.G. and S.I.S.; data curation, M.S.D., R.T., Y.F.E. and S.I.S.; formal analysis, M.S.D., R.T., Y.F.E., G.A.H., M.A.S., S.A.G., A.M.H. and S.I.S.; funding acquisition, S.A., A.M.H. and M.A.O.D.; investigation, M.S.D., S.A. and M.A.O.D.; methodology, M.S.D., R.T., Y.F.E., M.A.S., S.A.G. and S.I.S.; resources, M.S.D., R.T., Y.F.E., G.A.H. and S.I.S.; writing—original draft, G.A.H., M.A.O.D. and S.I.S.; writing—review and editing, A.M.H. All authors have read and agreed to the published version of the manuscript.

Funding: The work was funded by Taif University Researchers Supporting Project number (TURSP-2020/76), Taif University, Taif, Saudi Arabia.

Institutional Review Board Statement: The study was conducted according to the guidelines of the Faculty of Veterinary Medicine, Menufia University, Egypt (Directive 2010/63/EU).

Acknowledgments: The work was funded by Taif University Researchers Supporting Project number (TURSP-2020/76), Taif University, Taif, Saudi Arabia.

Conflicts of Interest: The authors declare no conflict of interest.

References

1. Hussein, J.; Schuberth, H.J. Recent advances in camel immunology. *Front. Immunol.* **2020**, *11*, 614150. [CrossRef]
2. Bekele, T.; Zeleke, M.; Baars, R. Milk production performance of the one-humped camel (*Camelus dromedarius*) under pastoral management in semi-arid eastern Ethiopia. *Livest. Prod. Sci.* **2002**, *76*, 37–44. [CrossRef]
3. Yagil, R. Cosmeceuticals: Camel and other milk—Natural skin maintenance. In *Complementary and Alternative Medicine: Breakthroughs in Research and Practice*; IGI Global: Hershey, PA, USA, 2019; pp. 95–124.
4. Bashir, M.E.A.A. Studies on Clinical, Aetiological and Antibiotic Susceptibility of Mastitis in She-Camel Camelus dromedarius in Butane Area, Sudan. Ph.D. Thesis, Sudan University of Science and Technology, Khartoum, Sudan, 2014.
5. Husein, A.; Haftu, B.; Hunde, A.; Tesfaye, A. Prevalence of camel (*Camelus dromedaries*) mastitis in jijiga town, Ethiopia. *Afr. J. Agric. Res.* **2013**, *8*, 3113–3120.
6. Jilo, K.; Galgalo, W.; Mata, W. Camel mastitis: A review. *MOJ Eco Environ. Sci.* **2017**, *2*, 00034. [CrossRef]
7. Shaheen, M.A.; Dapgh, A.N.; Kamel, E.; Ali, S.F.; Khairy, E.A.; Abuelhag, H.A.; Hakim, A.S. Advanced molecular characterization of enteropathogenic *Escherichia coli* isolated from diarrheic camel neonates in Egypt. *Vet. World* **2021**, *14*, 85–91. [CrossRef]

8. Abdulkadhim, M. Prevalence of methicillin resistance staphylococcus aureus in cattle and she-camels milk at Al-Qadisyia province. *Al-Anbar J. Vet. Sci.* **2012**, *5*, 63–67.

9. Abeer, A.; Gouda, A.; Dardir, H.; Ibrahim, A. Prevalence of some milk-borne bacterial pathogens threatening camel milk consumers in Egypt. *Global Vet* **2012**, *8*, 76–82.

10. Saleh, S.K.; Al-Ramadhan, G.; Faye, B. Monitoring of monthly SCC in she-camel in relation to milking practice, udder status and microbiological contamination of milk. *Emir. J. Food Agric.* **2013**, *25*, 403–408. [CrossRef]

11. Bessalah, S.; Fairbrother, J.M.; Salhi, I.; Vanier, G.; Khorchani, T.; Seddik, M.M.; Hammadi, M. Antimicrobial resistance and molecular characterization of virulence genes, phylogenetic groups of *Escherichia coli* isolated from diarrheic and healthy camel-calves in Tunisia. *Comp. Immunol. Microbiol. Infect. Dis.* **2016**, *49*, 1–7. [CrossRef] [PubMed]

12. Alhendi, A.B. Common diseases of camels (*Camelus dromedari*) in eastern province of Saudi Arabia. *Pak. Vet. J.* **2000**, *20*, 97–99.

13. Carvalho, I.; Tejedor-Junco, M.T.; Gonzalez-Martin, M.; Corbera, J.A.; Silva, V.; Igrejas, G.; Torres, C.; Poeta, P. *Escherichia coli* producing extended-spectrum beta-lactamases (es) from domestic camels in the canary islands: A one health approach. *Animals* **2020**, *10*, 1295. [CrossRef] [PubMed]

14. Gyles, C.L. Shiga toxin-producing *Escherichia coli*: An overview. *J. Anim. Sci.* **2007**, *85*, E45–E62. [CrossRef] [PubMed]

15. Kaper, J.B.; Karmali, M.A. The continuing evolution of a bacterial pathogen. *Proc. Natl. Acad. Sci. USA* **2008**, *105*, 4535–4536. [CrossRef] [PubMed]

16. Kaddu-Mulindwa, D.; Aisu, T.; Gleier, K.; Zimmermann, S.; Beutin, L. Occurrence of Shiga toxin-producing *Escherichia coli* in fecal samples from children with diarrhea and from healthy zebu cattle in Uganda. *Int. J. Food Microbiol.* **2001**, *66*, 95–101. [CrossRef]

17. FAO. *Shiga Toxin-Producing Escherichia coli (Stec) and Food: Attribution, Characterization and Monitoring*; World Health Organization: Geneva, Switzerland, 2019.

18. El-Sayed, A.; Ahmed, S.; Awad, W. Do camels (*Camelus dromedarius*) play an epidemiological role in the spread of Shiga toxin-producing *Escherichia coli* (stec) infection? *Trop. Anim. Health Prod.* **2008**, *40*, 469–473. [CrossRef]

19. Moore, J.; McCalmont, M.; Xu, J.; Nation, G.; Tinson, A.; Crothers, L.; Harron, D. Prevalence of faecal pathogens in calves of racing camels (*Camelus dromedarius*) in the United Arab emirates. *Trop. Anim. Health Prod.* **2002**, *34*, 283. [CrossRef] [PubMed]

20. Fadlelmula, A.; Al-Hamam, N.A.; Al-Dughaym, A.M. A potential camel reservoir for extended-spectrum beta-lactamase-producing *Escherichia coli* causing human infection in Saudi Arabia. *Trop. Anim. Health Prod.* **2016**, *48*, 427–433. [CrossRef] [PubMed]

21. Carvalho, I.; Silva, N.; Carroll, J.; Silva, V.; Currie, C.; Igrejas, G.; Poeta, P. Antibiotic resistance: Immunity-acquired resistance: Evolution of antimicrobial resistance among extended-spectrum β-lactamases and carbapenemases in Klebsiella pneumoniae and *Escherichia coli*. *Antibiot. Drug Resist.* **2019**, 239–259.

22. Diab, M.S.; Zaki, R.S.; Ibrahim, N.A.; El Hafez, M.S.A. Prevalence of multidrug resistance non-typhoidal salmonellae isolated from layer farms and humans in Egypt. *World Vet. J.* **2019**, *9*, 280–288. [CrossRef]

23. Clement, M.; Olabisi, M.; David, E.; Issa, M. Veterinary pharmaceuticals and antimicrobial resistance in developing countries. In *Veterinary Medicine and Pharmaceuticals*; IntechOpen: London, UK, 2019.

24. Erskine, R.; Walker, R.; Bolin, C.; Bartlett, P.; White, D. Trends in antibacterial susceptibility of mastitis pathogens during a seven-year period. *J. Dairy Sci.* **2002**, *85*, 1111–1118. [CrossRef]

25. Oliver, S.P.; Murinda, S.E.; Jayarao, B.M. Impact of antibiotic use in adult dairy cows on antimicrobial resistance of veterinary and human pathogens: A comprehensive review. *Foodborne Pathog. Dis.* **2011**, *8*, 337–355. [CrossRef]

26. Massé, J.; Lardé, H.; Fairbrother, J.M.; Roy, J.-P.; Francoz, D.; Dufour, S.; Archambault, M. Prevalence of antimicrobial resistance and characteristics of *Escherichia coli* isolates from fecal and manure pit samples on dairy farms in the province of Québec, Canada. *Front. Vet. Sci.* **2021**, *8*, 438. [CrossRef] [PubMed]

27. Dego, O.K. Current status of antimicrobial resistance and prospect for new vaccines against major bacterial bovine mastitis pathogens. In *Animal Reproduction in Veterinary Medicine*; IntechOpen: London, UK, 2020.

28. Seligsohn, D.; Nyman, A.K.; Younan, M.; Sake, W.; Persson, Y.; Bornstein, S.; Maichomo, M.; de Verdier, K.; Morrell, J.M.; Chenais, E. Subclinical mastitis in pastoralist dairy camel herds in Isiolo, Kenya: Prevalence, risk factors, and antimicrobial susceptibility. *J. Dairy Sci.* **2020**, *103*, 4717–4731. [CrossRef] [PubMed]

29. Mutua, J.; Gao, C.; Bebora, L.; Mutua, F. Antimicrobial resistance profiles of bacteria isolated from the nasal cavity of camels in Samburu, Nakuru, and Isiolo counties of Kenya. *J. Vet. Med.* **2017**, *2017*, 6. [CrossRef] [PubMed]

30. Suheir, I. Some Bacterial Species, Mycoplasma and Fungal Isolates Associated with Camel Mastitis. Master's Thesis, University of Khartoum, Khartoum, Sudan, 2004.

31. Miara, M.D.; Bendif, H.; Ait Hammou, M.; Teixidor-Tonneau, I. Ethnobotanical survey of medicinal plants used by nomadic peoples in the Algerian steppe. *J. Ethnopharmacol.* **2018**, *219*, 248–256. [CrossRef] [PubMed]

32. Owusu, E.; Ahorlu, M.M.; Afutu, E.; Akumwena, A.; Asare, G.A. Antimicrobial activity of selected medicinal plants from a sub-Saharan African country against bacterial pathogens from post-operative wound infections. *Med. Sci.* **2021**, *9*, 23.

33. Aziz, M.A.; Khan, A.H.; Adnan, M.; Ullah, H. Traditional uses of medicinal plants used by indigenous communities for veterinary practices at Bajaur Agency, Pakistan. *J. Ethnobiol. Ethnomed.* **2018**, *14*, 1–18. [CrossRef] [PubMed]

34. CLSI. *Performance Standards for Antimicrobial Susceptibility Testing; Twenty-Fifth Informational Supplement*; CLSI document M100-S25; Clinical and Laboratory Standards Institute: Wayne, PA, USA, 2015.

35. Balemi, A.; Gumi, B.; Amenu, K.; Girma, S.; Gebru, M.; Tekle, M.; Ríus, A.A.; D'Souza, D.H.; Agga, G.E.; Kerro Diego, O. Prevalence of mastitis and antibiotic resistance of bacterial isolates from CMT positive milk samples obtained from dairy cows, camels, and goats in two pastoral districts in southern Ethiopia. *Animals* **2021**, *11*, 1530. [CrossRef]
36. Alizade, H.; Hosseini Teshnizi, S.; Azad, M.; Shoji, S.; Gouklani, H.; Davoodian, P.; Ghanbarpour, R. An overview of diarrheagenic *Escherichia coli* in Iran: A systematic review and meta-analysis. *J. Res. Med. Sci.* **2019**, *24*, 23.
37. Galal, H.; Hakim, A.; Dorgham, S. Phenotypic and virulence genes screening of *Escherichia coli* strains isolated from different sources in delta Egypt. *Life Sci. J.* **2013**, *10*, 352–361.
38. Osman, K.M.; Mustafa, A.M.; Elhariri, M.; Abdelhamid, G.S. The distribution of *Escherichia coli* serovars, virulence genes, gene association and combinations and virulence genes encoding serotypes in pathogenic e. Coli recovered from diarrhoeic calves, sheep and goat. *Transbound Emerg. Dis.* **2013**, *60*, 69–78. [CrossRef]
39. Hakim, A.S.; Omara, S.T.; Same, S.M.; Fouad, E.A. Serotyping, antibiotic susceptibility, and virulence genes screening of *Escherichia coli* isolates obtained from diarrheic buffalo calves in Egyptian farms. *Vet. World* **2017**, *10*, 769. [CrossRef]
40. Abo Hashem, M.; Ibrahim, S.; Goda, A.S.; Enany, M. Diversity of microorganisms associated to she camels' subclinical and clinical mastitis in South Sinai, Egypt. *Suez Canal Vet. Med. J. SCVMJ* **2020**, *25*, 307–319. [CrossRef]
41. Al Humam, N.A. Special biochemical profiles of *Escherichia coli* strains isolated from humans and camels by the Vitek 2 automated system in al-ahsa, Saudi Arabia. *Afr. J. Microbiol. Res.* **2016**, *10*, 783–790.
42. El-Hewairy, H.; Awad, W.; Ibrahim, A. Serotyping and molecular characterization of *Escherichia coli* isolated from diarrheic and in-contact camel calves. *Egypt. J. Comp. Pathol. Clin. Pathol.* **2009**, *22*, 216–233.
43. Al-Ajmi, D.; Rahman, S.; Banu, S. Occurrence, virulence genes, and antimicrobial profiles of *Escherichia coli* o157 isolated from ruminants slaughtered in al ain, united Arab emirates. *BMC Microbiol.* **2020**, *20*, 210. [CrossRef]
44. Mohammed, H.O.; Stipetic, K.; Salem, A.; McDonough, P.; Chang, Y.F.; Sultan, A. Risk of *Escherichia coli* o157: H7, non-o157 Shiga toxin-producing *Escherichia coli*, and campylobacter spp. In food animals and their products in Qatar. *J. Food Prot.* **2015**, *78*, 1812–1818. [CrossRef]
45. Baschera, M.; Cernela, N.; Stevens, M.J.A.; Liljander, A.; Jones, J.; Corman, V.M.; Nuesch-Inderbinen, M.; Stephan, R. Shiga toxin-producing *Escherichia coli* (stec) isolated from fecal samples of African dromedary camels. *One Health* **2019**, *7*, 100087. [CrossRef] [PubMed]
46. Adamu, M.S.; Ugochukwu, I.C.I.; Idoko, S.I.; Kwabugge, Y.A.; Abubakar, N.S.; Ameh, J.A. Virulent gene profile and antibiotic susceptibility pattern of Shiga toxin-producing *Escherichia coli* (stec) from cattle and camels in Maiduguri, north-eastern Nigeria. *Trop. Anim. Health Prod.* **2018**, *50*, 1327–1341. [CrossRef] [PubMed]
47. EL-Alfy, S.M.; Ahmed, S.F.; Selim, S.A.; Aziz, M.H.A.; Zakaria, A.M.; Klena, J.D. Prevalence and characterization of Shiga toxin o157 and non-o157 enterohemorrhagic *Escherichia coli* isolated from different sources in Ismailia, Egypt. *Afr. J. Microbiol. Res.* **2013**, *7*, 2637–2645.
48. Ramadan, H.; Awad, A.; Ateya, A. Detection of phenotypes, virulence genes and phylotypes of avian pathogenic and human diarrheagenic *Escherichia coli* in Egypt. *J. Infect. Dev. Ctries* **2016**, *10*, 584–591. [CrossRef] [PubMed]
49. Persson, S.; Olsen, K.E.; Scheutz, F.; Krogfelt, K.A.; Gerner-Smidt, P. A method for fast and simple detection of major diarrhoeagenic *Escherichia coli* in the routine diagnostic laboratory. *Clin. Microbiol. Infect.* **2007**, *13*, 516–524. [CrossRef]
50. Monaghan, A.; Byrne, B.; Fanning, S.; Sweeney, T.; McDowell, D.; Bolton, D.J. Serotypes and virulence profiles of non-o157 Shiga toxin-producing *Escherichia coli* isolates from bovine farms. *Appl. Environ. Microbiol.* **2011**, *77*, 8662–8668. [CrossRef]
51. Moses, A.; Egwu, G.; Ameh, J. Antimicrobial-resistant pattern of e. Coli o157 isolated from human, cattle and surface water samples in northeast Nigeria. *J. Vet. Adv.* **2012**, *2*, 209–215.
52. Bakhtiari, N.M.; Fazlara, A.; Jorge, H. Risk of Shiga toxin-producing *Escherichia coli* infection in humans due to consuming unpasteurized dairy products. *Int. J. Enteric. Pathog.* **2018**, *6*, 14–17. [CrossRef]
53. Yamasaki, E.; Watahiki, M.; Isobe, J.; Sata, T.; Nair, G.B.; Kurazono, H. Quantitative detection of Shiga toxins directly from stool specimens of patients associated with an outbreak of enterohemorrhagic *Escherichia coli* in Japan—Quantitative Shiga toxin detection from stool during ehec outbreak. *Toxins* **2015**, *7*, 4381–4389. [CrossRef] [PubMed]
54. Sethulekshmi, C.; Latha, C.; Anu, C. Occurrence and Quantification of Shiga Toxin-Producing Escherichia coli from Food Matrices. *Vet. World* **2018**, *11*, 104–111. [CrossRef] [PubMed]
55. Angulo, F.J.; Steinmuller, N.; Demma, L.; Bender, J.B.; Eidson, M.; Angulo, F.J. Outbreaks of enteric disease associated with animal contact: Not just a foodborne problem anymore. *Clin. Infect. Dis.* **2006**, *43*, 1596–1602. [CrossRef]
56. Ranjbar, R.; Masoudimanesh, M.; Dehkordi, F.S.; Jonaidi-Jafari, N.; Rahimi, E. Shiga (Vero)-toxin producing *Escherichia coli* isolated from the hospital foods; virulence factors, o-serogroups and antimicrobial resistance properties. *Antimicrob. Resist. Infect. Control* **2017**, *6*, 4. [CrossRef]
57. Rashid, M.; Kotwal, S.K.; Malik, M.; Singh, M. Prevalence, genetic profile of virulence determinants and multidrug resistance of *Escherichia coli* isolates from foods of animal origin. *Vet. World* **2013**, *6*, 139–142. [CrossRef]
58. Mashak, Z. Virulence genes and phenotypic evaluation of the antibiotic resistance of Vero toxin-producing *Escherichia coli* recovered from milk, meat, and vegetables. *Jundishapur J. Microbiol.* **2018**, *11*, e62288. [CrossRef]
59. Aly, S.M.; Albutti, A. Antimicrobials use in aquaculture and their public health impact. *J. Aquac. Res. Dev.* **2014**, *5*, 1.
60. Shrivastava, S.R.; Shrivastava, P.S.; Ramasamy, J. World health organization releases global priority list of antibiotic-resistant bacteria to guide research, discovery, and development of new antibiotics. *J. Med. Soc.* **2018**, *32*, 76. [CrossRef]

61. Diab, M.S.; Ibrahim, N.A.; Elnaker, Y.F.; Zidan, S.A.; Saad, M.A. Molecular detection of staphylococcus aureus enterotoxin genes isolated from mastitic milk and humans in el-behira, Egypt. *Int. J. One Health* **2021**, *7*, 70–77. [CrossRef]
62. Abdo, S.E.; Gewaily, M.S.; Abo-Al-Ela, H.G.; Almere, R.; Soliman, A.A.; Elkomy, A.H.; Dawood, M.A.O. Vitamin c rescues inflammation, immunosuppression, and histopathological alterations induced by chlorpyrifos in Nile tilapia. *Environ. Sci. Pollut. Res.* **2021**, *28*, 28750–28763. [CrossRef]
63. Wenz, J.; Barrington, G.; Garry, F.; Ellis, R.; Magnuson, R. *Escherichia coli* isolates' serotypes, genotypes, and virulence genes and clinical coliform mastitis severity. *J. Dairy Sci.* **2006**, *89*, 3408–3412. [CrossRef]
64. De Verdier, K.; Nyman, A.; Greko, C.; Bengtsson, B. Antimicrobial resistance and virulence factors in *Escherichia coli* from Swedish dairy calves. *Acta Vet. Scand.* **2012**, *54*, 2. [CrossRef]
65. Scaria, J.; Warnick, L.D.; Kaneene, J.B.; May, K.; Teng, C.H.; Chang, Y.F. Comparison of phenotypic and genotypic antimicrobial profiles in *Escherichia coli* and salmonella enterica from the same dairy cattle farms. *Mol. Cell Probes* **2010**, *24*, 325–345. [CrossRef] [PubMed]
66. Momtaz, H.; Farzan, R.; Rahimi, E.; Safarpoor Dehkordi, F.; Sound, N. Molecular characterization of Shiga toxin-producing *Escherichia coli* isolated from ruminant and donkey raw milk samples and traditional dairy products in Iran. *Sci. World J.* **2012**, *2012*, 231342. [CrossRef]
67. Ranjbar, R.; Safarpoor Dehkordi, F.; Sakhaei Shahreza, M.H.; Rahimi, E. Prevalence, identification of virulence factors, o-serogroups and antibiotic resistance properties of Shiga-toxin-producing *Escherichia coli* strains isolated from raw milk and traditional dairy products. *Antimicrob. Resist. Infect. Control* **2018**, *7*, 53. [CrossRef]
68. Ababu, A.; Endashaw, D.; Fesseha, H. Isolation and antimicrobial susceptibility profile of *Escherichia coli* o157: H7 from raw milk of dairy cattle in holeta district, central Ethiopia. *Int. J. Microbiol.* **2020**, *2020*, 6626488. [CrossRef]
69. Hoque, M.N.; Das, Z.C.; Talukder, A.K.; Alam, M.S.; Rahman, A.N. Different screening tests and milk somatic cell count for the prevalence of subclinical bovine mastitis in Bangladesh. *Trop. Anim. Health Prod.* **2015**, *47*, 79–86. [CrossRef] [PubMed]
70. Quinn, P.J.; Markey, B.K.; Leonard, F.C.; Hartigan, P.; Fanning, S.; Fitzpatrick, E. *Veterinary Microbiology and Microbial Disease*; John Wiley & Sons: Hoboken, NJ, USA, 2011.
71. Kok, T.; Worswich, D.; Gowans, E. Some serological techniques for microbial and viral infections. In *Practical Medical Microbiology*; Collee, J., Fraser, A., Marmion, B., Simmons, A., Eds.; Churchill Livingstone: Edinburgh, UK, 1996.
72. Paton, J.C.; Paton, A.W. Pathogenesis and diagnosis of Shiga toxin-producing *Escherichia coli* infections. *Clin. Microbiol. Rev.* **1998**, *11*, 450–479. [CrossRef] [PubMed]

Article

Antimicrobial Resistance Genes and Diversity of Clones among Faecal ESBL-Producing *Escherichia coli* Isolated from Healthy and Sick Dogs Living in Portugal

Isabel Carvalho [1,2,3,4,5], Rita Cunha [6], Carla Martins [7], Sandra Martínez-Álvarez [5], Nadia Safia Chenouf [5,8], Paulo Pimenta [9], Ana Raquel Pereira [10], Sónia Ramos [11], Madjid Sadi [5,12], Ângela Martins [13], Jorge Façanha [14], Fazle Rabbi [15], Rosa Capita [16,17], Carlos Alonso-Calleja [16,17], Maria de Lurdes Nunes Enes Dapkevicius [18,19], Gilberto Igrejas [2,3,4], Carmen Torres [5] and Patrícia Poeta [1,4,*]

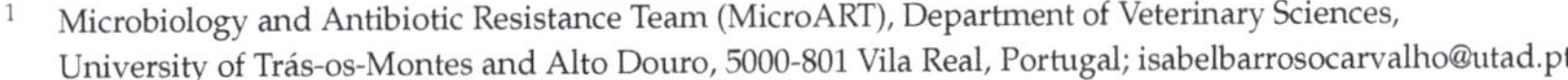

1. Microbiology and Antibiotic Resistance Team (MicroART), Department of Veterinary Sciences, University of Trás-os-Montes and Alto Douro, 5000-801 Vila Real, Portugal; isabelbarrosocarvalho@utad.pt
2. Department of Genetics and Biotechnology, University of Trás-os-Montes and Alto Douro, 5000-801 Vila Real, Portugal; gigrejas@utad.pt
3. Functional Genomics and Proteomics Unit, University of Trás-os-Montes and Alto Douro, 5000-801 Vila Real, Portugal
4. Laboratory Associated for Green Chemistry (LAQV-REQUIMTE), New University of Lisbon, 2829-516 Monte da Caparica, Portugal
5. Area Biochemistry and Molecular Biology, University of La Rioja, 26006 Logroño, Spain; sandra.martinezal@alum.unirioja.es (S.M.-Á.); chenoufns@gmail.com (N.S.C.); sadimadjid@gmail.com (M.S.); carmen.torres@unirioja.es (C.T.)
6. Hospital Veterinário Cascais da Onevet, 2775-352 Parede, Lisbon, Portugal; ritalaurentinocunha@gmail.com
7. Clínica Veterinária do Vouga, 3740-253 Sever do Vouga, Portugal; clinica.vouga@gmail.com
8. Laboratory of Exploration and Valuation of the Steppe Ecosystem, University of Djelfa, Djelfa 17000, Algeria
9. Hospital Veterinário de Trás-os-Montes, 5000-056 Vila Real, Portugal; paulo.pimenta@onevetgroup.pt
10. Centro Veterinário de Macedo de Cavaleiros, 5340-202 Bragança, Portugal; vetcenterbrg@gmail.com
11. VetRedondo, Consultório Veterinário de Monte Redondo Unipessoal Lda, Monte Redondo, 2425-618 Leiria, Portugal; soniacatarinaramos@gmail.com
12. Laboratory of Biotechnology Related to Animals Reproduction, Université Saad Dahlab de Blida, Blida 09000, Algeria
13. Animal and Veterinary Research Center (CECAV), University of Trás-os-Montes and Alto Douro, 5000-801 Vila Real, Portugal; angela@utad.pt
14. Centro Veterinário Jorge Façanha, 5140-060 Carrazeda de Ansiães, Portugal; jorgefacanha@hotmail.com
15. Australian Computer Society, Docklands, Melbourne, VIC 3008, Australia; bilirabbi@yahoo.com
16. Department of Food Hygiene and Technology, Veterinary Faculty, University of León, 24071 León, Spain; rosa.capita@unileon.es (R.C.); carlos.alonso.calleja@unileon.es (C.A.-C.)
17. Institute of Food Science and Technology, University of León, 24071 León, Spain
18. Faculty of Agricultural and Environmental Sciences, University of the Azores, 9500-321 Angra do Heroísmo, Portugal; maria.ln.dapkevicius@uac.pt
19. Institute of Agricultural and Environmental Research and Technology (IITAA), University of the Azores, 9500-321 Angra do Heroísmo, Portugal
* Correspondence: ppoeta@utad.pt; Tel.: +351-25935-0466; Fax: +351-25935-0629

check for updates

Citation: Carvalho, I.; Cunha, R.; Martins, C.; Martínez-Álvarez, S.; Safia Chenouf, N.; Pimenta, P.; Pereira, A.R.; Ramos, S.; Sadi, M.; Martins, Â.; et al. Antimicrobial Resistance Genes and Diversity of Clones among Faecal ESBL-Producing *Escherichia coli* Isolated from Healthy and Sick Dogs Living in Portugal. *Antibiotics* **2021**, *10*, 1013. https://doi.org/10.3390/antibiotics10081013

Academic Editors: Elisabete Silva and Manuela Oliveira

Received: 26 June 2021
Accepted: 16 August 2021
Published: 20 August 2021

Publisher's Note: MDPI stays neutral with regard to jurisdictional claims in published maps and institutional affiliations.

Abstract: The purpose of this study was to analyse the prevalence and genetic characteristics of ESBL and acquired-AmpC (qAmpC)-producing *Escherichia coli* isolates from healthy and sick dogs in Portugal. Three hundred and sixty-one faecal samples from sick and healthy dogs were seeded on MacConkey agar supplemented with cefotaxime (2 μg/mL) for cefotaxime-resistant (CTX^R) *E. coli* recovery. Antimicrobial susceptibility testing for 15 antibiotics was performed and the ESBL-phenotype of the *E. coli* isolates was screened. Detection of antimicrobial resistance and virulence genes, and molecular typing of the isolates (phylogroups, multilocus-sequence-typing, and specific-ST131) were performed by PCR (and sequencing when required). CTX^R *E. coli* isolates were obtained in 51/361 faecal samples analysed (14.1%), originating from 36/234 sick dogs and 15/127 healthy dogs. Forty-seven ESBL-producing *E. coli* isolates were recovered from 32 sick (13.7%) and 15 healthy animals (11.8%). Different variants of *bla*CTX-M genes were detected among 45/47 ESBL-producers: *bla*CTX-M-15 ($n = 26$), *bla*CTX-M-1 ($n = 10$), *bla*CTX-M-32 ($n = 3$), *bla*CTX-M-55 ($n = 3$), *bla*CTX-M-14 ($n = 2$), and *bla*CTX-M-variant ($n = 1$); one ESBL-positive isolate co-produced CTX-M-15 and CMY-2 enzymes. Moreover, two additional CTX^R ESBL-negative *E. coli* isolates were

CMY-2-producers (qAmpC). Ten different sequence types were identified (ST/phylogenetic-group/β-lactamase): ST131/B2/CTX-M-15, ST617/A/CTX-M-55, ST3078/B1/CTX-M-32, ST542/A/CTX-M-14, ST57/D/CTX-M-1, ST12/B2/CTX-M-15, ST6448/B1/CTX-M-15 + CMY-2, ST5766/A/CTX-M-32, ST115/D/CMY-2 and a new-ST/D/CMY-2. Five variants of CTX-M enzymes (CTX-M-15 and CTX-M-1 predominant) and eight different clonal complexes were detected from canine ESBL-producing *E. coli* isolates. Although at a lower rate, CMY-2 β-lactamase was also found. Dogs remain frequent carriers of ESBL and/or qAmpC-producing *E. coli* with a potential zoonotic role.

Keywords: antimicrobial resistance; dogs; *Escherichia coli*; ESBL; CTX-M-15; CTX-M-1; CTX-M-32; CTX-M-55; CTX-M-14; qAmpC; CMY-2

---

1. Introduction

Antimicrobial resistance has become a major challenge for public health worldwide. The selective pressure, which results from the long-term use of antibiotics, allowed bacterial species to be resistant to these agents. It has been believed that this resistance is reaching alarming levels, considering that resistance rates have risen extremely, during the last two decades [1,2].

Escherichia coli, a Gram-negative bacterium belonging to the *Enterobacteriaceae* family, is a common member of the intestinal microbiota of humans and companion animals [3,4]. However, this opportunistic pathogen can cause intestinal and extra-intestinal diseases. It may contribute, in many cases, to antimicrobial resistance dissemination. Recently, the World Health Organization [5] published a global priority list of antibiotic-resistant bacteria, where third-generation cephalosporin- and/or carbapenem-resistant *Enterobacteriaceae*, including *E. coli*, were included in the Priority 1 group. It is important to note that first-generation cephalosporins and amoxicillin + clavulanic acid are among the most prescribed drugs for dogs [3,4,6].

During recent years, the emergence and rapid dissemination of *Enterobacteriaceae* carrying genes encoding the extended-spectrum-β-lactamases (ESBLs), acquired AmpC β-lactamases (qAmpC), or carbapenemases are considered of great concern [4,7]. One of the most important mechanisms is the plasmid-mediated production of extended-spectrum β-lactamases (ESBLs), which can hydrolyse broad-spectrum cephalosporins (such as cefotaxime). The horizontal gene transfer (HGT) among bacteria is driven by plasmids [8,9], which play an important role in the transference of antibiotic-resistance genes among bacteria, contributing to the spread of multidrug resistance (MDR), and limiting therapeutic options [10]. ESBLs of the CTX-M-type and the qAmpC CMY-2 are increasingly being reported in bacteria worldwide, while livestock or companion animals are potential sources, leading to the spread of β-lactam-resistant bacteria in humans [11,12].

The close proximity between dogs and their owners increases the possibility of transmitting resistant bacteria [13,14]. According to Dupouy et al. [6], dogs could transmit MDR bacteria due to their close contact with humans, the high consumption of β-lactams in small animal veterinary practice, and also the frequent occurrence of ESBL/qAmpC-producing *E. coli*. The occurrence of ESBL-producing *E. coli* has been widely reported in both healthy companion animals [12,15] and diseased ones [1,16–19]. International high-risk clones of *E. coli* are frequently detected worldwide, not only in human infections but also in those of companion animals [2,3,17]. Over the past 5 years, the presence of ESBL/qAmpC genes in *Enterobacteriaceae* strains from faeces of dogs in Europe has been reported in several studies [6,12,13,20], including Portugal [21,22]. However, knowledge about the clonality of ESBL/qAmpC-producing isolates and the potential zoonotic reservoir of human-associated STs is not well documented. Moreover, there is still a lack of data about their prevalence in sick and healthy dogs, simultaneously. In this study, we aim at characterizing the prevalence and diversity of ESBL- and qAmpC- producing *E. coli* faecal isolates from healthy and sick dogs in Portugal, as well as determining their genetic lineages and phylogenetic groups.

2. Materials and Methods

2.1. Animals and Sampling

A total of 361 faecal samples were recovered from 127 healthy and 234 hospitalized dogs from different cities in Portugal. All samples were collected between April andAugust 2017 (one sample/animal) using standardized procedures [23].

The hospitalized dogs came from 7 different veterinary hospitals or clinic centers; the healthy dogs came from a local kennel located in Vila Real ($n = 31$) and from local houses ($n = 96$). The seven hospitals/clinic centers were located in different centers of the Portuguese territory: Bragança (1 hospital, $n = 29$ dogs), Vila Real (4 hospitals, $n = 62$), Aveiro (1 hospital, $n = 58$), Leiria (1 hospital, $n = 17$), and Lisbon (1 hospital, $n = 68$) (Figure S1). It is important to note that faecal samples from unhealthy dogs were collected from the ordinary population of animals hospitalized in hospitals or veterinary clinics, not endangering their health, or causing harm or pain. In the same line, faecal samples from healthy animals were also recovered by their owners. All of them were analysed with the owner's permission or with kennel collaboration. The faecal samples were dispatched immediately to the Microbiology Laboratory of the University of Trás-os-Montes and Alto-Douro (UTAD).

2.2. E. coli Isolation

From each faecal sample, a small portion of 2 g was diluted in Brain Heart Infusion (BHI, Condalab, Spain) and incubated in aerobic conditions for 24 h at 37 °C. After that, samples were seeded on MacConkey agar (Becton, Dickinson and Company Sparks, Le Pont de Claix, France) supplemented with cefotaxime (2 µg/mL) and incubated for 24 h at 37 °C. Colonies showing *E. coli* morphology were recovered (one colony per sample) and identified by a classical biochemical method named IMViC (Indol, Methyl-red, Voges–Proskauer, and Citrate).

The matrix-assisted laser desorption/ionization time-of-flight mass spectrometry method (MALDI-TOF MS, MALDI Biotyper®from Bruker Daltonik, Bremen, Germany) was applied in this study to confirm bacterial species identification. *E. coli* isolates were kept at −80 °C and were further characterized.

2.3. Susceptibility Testing

Antimicrobial susceptibility testing was performed using the Kirby–Bauer disk diffusion method and according to Clinical and Laboratory Standards Institute guidelines (2019) [24] for the following 15 antibiotics (µg/disk): ampicillin (10), amoxicillin + clavulanic acid (20), cefotaxime (30), cefoxitin (30), ceftazidime (30), aztreonam (30), imipenem (10), gentamicin (10), streptomycin (10), ciprofloxacin (5), trimethoprim-sulfamethoxazole (1.25 ± 23.75), amikacin (30), tobramycin (10), tetracycline (30), and chloramphenicol (30). In addition, the screening of phenotypic ESBL production was carried out by the double-disk synergy test using cefotaxime, ceftazidime, and amoxicillin/clavulanic discs in Mueller Hinton (MH) agar (Condalab, Spain) [24].

2.4. DNA Extraction and Quantification

Genomic DNA from cefotaxime-resistant (CTX^R) isolates were extracted using the boiled method [25]. In order to quantify the nucleic acid concentration and the level of purity, the absorbance readings were taken at 260 and 280 nm (Spectrophotometer ND-100, Nanodrop, Thermo Fisher Scientific, Waltham, MA USA).

2.5. Antibiotic Resistance and Virulence Genes Detection

The genetic basis of resistance was investigated using PCR methods and subsequent sequencing of the obtained amplicons (specific genes). Negative and positive controls of the University of La Rioja were used in this work. Moreover, the data regarding PCR conditions for each primer (Sigma-Aldrich, Madrid, Spain) as well as the size of the obtained amplicons that were sequenced are illustrated in detail in Table S1.

The presence of bla_{CTX-M} (Groups 1 and 9), bla_{CMY-2}, bla_{DHA-1}, bla_{TEM}, bla_{SHV}, bla_{VEB}, $bla_{KPC2/3}$, bla_{NDM}, bla_{OXA-48}, and bla_{VIM} was tested by PCR/sequencing (Table S1) [26–30]. Furthermore, the *mcr*-1 gene (colistin resistance) [31], *tet*A/*tet*B (tetracycline resistance) [32], $stx_{1,2}$ genes related to Shiga toxin-producing *E. coli* (STEC) [33], and *int*1 gene (integrase of class 1 integrons) and its variable region (RV *int*1) were also analysed by PCR/sequencing [30]. Analysis of DNA sequences was performed using the standard databases (nucleotide collection) in the BLASTN program (2021 version), available at the National Center for Biotechnology Information (https://blast.ncbi.nlm.nih.gov/Blast.cgi (accessed on 31 January 2021).

2.6. Multilocus Sequence Typing and Phylogroup Typing of E. coli Isolates

Multilocus sequence typing (MLST), by the analysis of seven housekeeping genes (*fum*C, *adk*, *pur*A, *icd*, *rec*A, *mdh*, and *gyr*B), was carried out for thirteen representative *E. coli* isolates (based on the antimicrobial resistance phenotype) according to the protocol described on PubMLST (Public databases for molecular typing and microbial genome diversity) website [34]. The allele combination was determined after sequencing of the seven genes, and the sequence type (ST) and clonal complex (CC) were identified.

Phylogenetic classification of all *E. coli* isolates was performed according to the presence of *chu*A, *yja*A, and TSPE4.C2 genes [35].

2.7. Statistical Analyses

All statistical analyses were performed using the JMP Statistics software (v7.0, SAS Institute). The Pearson's Chi-square and Fisher's exact tests were performed to understand and identify the associations between the origin of strain (healthy or sick dog) and antibiotic resistance (antibiotic and gene). In this line, we consider two categorical variables: the sick or healthy animal, and the resistance for each antibiotic/gene. A p-value < 0.05 was established as indicating statistical significance [36].

3. Results

CTX^R *E. coli* isolates were recovered in 51/361 faecal samples tested (14.1%), originating from 36/234 sick dogs (15.4%) and 15/127 healthy dogs (11.8%). These CTX^R isolates were detected among 29 male dogs (56.9%) and 22 female dogs (43.1%); most of them belonged to an undetermined breed ($n = 38$), followed by the Labrador/Golden Retriever breed ($n = 4$), while the remaining dogs belonged to different pure breeds (Tables 1 and 2).

Forty-seven ESBL-producing *E. coli* isolates were detected among the 51 CTX^R isolates, recovered from 32 sick and 15 healthy dogs (frequencies of 13.7% and 11.8%, respectively). The phenotypes of antibiotic resistance for these ESBL-producing isolates are shown in Table 1 and the rates of antibiotic resistance of these isolates depending on their origin (sick or healthy dogs) are represented in Figure 1. No statistical difference could be established between the origin of the strain (healthy or sick dog) and the resistance to different antibiotics ($p > 0.05$) (Figure 1).

The two remaining ESBL-positive isolates were revealed negative to all ESBL genes under study. Furthermore, a bla_{TEM} gene was detected in eight bla_{CTX-M}-producing isolates. On the other hand, six ESBL-positive isolates showed cefoxitin-resistance (FOX^R), and the bla_{CMY-2} gene was detected in one CTX-M-15-producing isolate obtained from a sick dog; the others ESBL-positive-FOX^R isolates were negative for bla_{CMY-2} and bla_{DHA} genes by PCR. Among the ESBL-positive isolates, resistance to tetracycline was mediated by the *tet*A (24 isolates) and/or *tet*B genes (Table 1).

Table 1. Phenotypic and molecular features of the 47 ESBL-producing *E. coli* isolates recovered from healthy and sick dogs in Portugal.

Isolate Number	Origin [a]	Sick/Healthy	Gender [b]	Age [c]	Breed [d]	Phenotype of Antibiotic Resistance [e]	β-Lactamases	Other Genes and Integrons [f]	PG [g]	MLST [h]
X605	HV Lisboa	Sick	M	15A	UD	AMP, CTX, ATM, CHL, CIP, TET	CTX-M-15	*tet*(A)	B1	ST6448
X614	HV Lisboa	Sick	F	2A	UD	AMP, AUG, FOX, CTX, CAZ, ATM, CHL, CIP, TET	CTX-M-15, CMY-2	*tet*(A)	B1	ST6448
X607	HV Lisboa	Sick	F	1A	UD	AMP, AUG, CTX, CAZ, ATM, CIP, TOB, CN, S, TET	CTX-M-15	*tet*(A)	B2	ST131
X610	HV Lisboa	Sick	F	1,5A	UD	AMP, CTX, CAZ, ATM, CIP, S, TET	CTX-M-15	*tet*(A)	B2	ST12
X603	CV Bragança	Sick	F	10A	UD	AMP, AUG, CTX, CAZ, ATM, CIP, TOB, CN, S, TET	CTX-M-15	*tet*(A)	B2	ST131
X602	CV VR	Sick	F	3A	UD	AMP, AUG, CTX, CHL, TOB, CN, S, TET	CTX-M-15	*tet*(A)	B2	ST12
X558	Kennel	Healthy	M	2A	Labrador	AMP, AUG, CTX, CAZ, CIP, SXT, S, TET	CTX-M-15, TEM	*int*1	B1	NT
X562	Kennel	Healthy	F	4M	UD	AMP, CTX, CIP, SXT, S, TET	CTX-M-15, TEM	*tet*(A), *int*1	B1	NT
X569	HVTM	Sick	M	4A	UD	AMP, AUG, CTX, CAZ, ATM, CIP, SXT, TOB, CN, S, TET	CTX-M-15, TEM	*tet*(A), *tet*(B), *int*1	A	NT
X575	CV Transm	Sick	M	4A	UD	AMP, AUG, CTX, CAZ, ATM, CIP, SXT, TOB, CN, TET	CTX-M-15, TEM	*tet*(A), *tet*(B), *int*1	A	NT
C10151	HV Lisboa	Sick	F	11A	UD	AMP, AUG, CTX, TET	CTX-M-15, TEM	ND	A	NT
X550	HD	Healthy	F	8A	UD	AMP, AUG, CTX, CAZ, ATM, CHL, CIP, SXT, TET	CTX-M-15	*tet*(A), *int*1	B1	NT
X556	HD	Healthy	F	14A	Yorkshire	AMP, AUG, FOX, CTX, ATM	CTX-M-15	ND	B1	NT
X563	Kennel	Healthy	M	5A	Rottweiler	AMP, AUG, CTX, TET	CTX-M-15	ND	A	NT
X588	HVTM	Sick	M	5A	UD	AMP, AUG, CTX, CAZ, ATM, CHL, CN, TET	CTX-M-15	*tet*(A)	D	NT

Table 1. *Cont.*

Isolate Number	Origin [a]	Sick/Healthy	Gender [b]	Age [c]	Breed [d]	Phenotype of Antibiotic Resistance [e]	β-Lactamases	Other Genes and Integrons [f]	PG [g]	MLST [h]
X598	HVTM	Sick	M	6A	Russell Terrier	AMP, AUG, CTX, CAZ, ATM, CHL, CIP, SXT, TET	CTX-M-15	ND	B1	NT
X576	HV Lisboa	Sick	M	15A	UD	AMP, CTX, CAZ, ATM, CHL, CIP, SXT, TET	CTX-M-15	*tet*(A), *int*1	D	NT
X577	HV Lisboa	Sick	F	6M	UD	AMP, AUG, CTX, CAZ, ATM, CHL, CIP, SXT, TET	CTX-M-15	*tet*(A), *int*1	B1	NT
X578	HV Lisboa	Sick	M	13A	UD	AMP, AUG, FOX, CTX, CAZ, ATM, CHL, CIP, SXT, TOB, TET	CTX-M-15	*tet*(A)	B1	NT
X580	HV Lisboa	Sick	F	5A	UD	AMP, CTX, CAZ, ATM, CHL, CIP, SXT, TET	CTX-M-15	*int*1	B1	NT
X584	HV Lisboa	Sick	M	5A	UD	AMP, CTX, CAZ, ATM, CHL, CIP, SXT, TET	CTX-M-15	*tet*(A), *int*1	B1	NT
X604	HV Lisboa	Sick	M	12A	UD	AMP, AUG, CTX, ATM	CTX-M-15	ND	D	NT
X618	HV Lisboa	Sick	F	2A	UD	AMP, AUG, FOX, CTX, CAZ, ATM, CHL, CIP, SXT, TET	CTX-M-15	*tet*(A), *int*1	B1	NT
X620	HV Lisboa	Sick	M	9A	UD	AMP, AUG, CTX, CAZ, ATM, CIP, SXT, TOB, CN, S, TET	CTX-M-15	*tet*(A), *int*1	D	NT
X622	HV Lisboa	Sick	M	3A	UD	AMP, AUG, FOX, CTX, CAZ, ATM, CIP, SXT, S, TET	CTX-M-15	ND	A	NT
X599	CV Bragança	Sick	M	7A	Rodengo	AMP, AUG, FOX, CTX, CAZ, ATM, CHL, CIP, TET	CTX-M-15	ND	B1	NT
C10264	CV Vouga	Sick	F	9M	Pincher	AMP, CTX, CAZ	CTX-M-1	ND	D	ST57
X554	HVTM	Sick	M	1A	Labrador	AMP, CTX, CAZ, TET	CTX-M-1, TEM	ND	A	NT
X557	Kennel	Healthy	F	3A	Serra Estrela	AMP, AUG, CTX, CAZ, TOB, AK, S	CTX-M-1	ND	B1	NT
X559	Kennel	Healthy	F	3M	UD	AMP, AUG, CTX, CAZ	CTX-M-1	ND	D	NT
X560	Kennel	Healthy	M	1A	Labrador	AMP, AUG, CTX, CAZ, TET	CTX-M-1	ND	B1	NT

Table 1. *Cont.*

Isolate Number	Origin [a]	Sick/Healthy	Gender [b]	Age [c]	Breed [d]	Phenotype of Antibiotic Resistance [e]	β-Lactamases	Other Genes and Integrons [f]	PG [g]	MLST [h]
X581	HV Lisboa	Sick	M	14A	UD	AMP, AUG, CTX, CAZ, TET	CTX-M-1	*tet*(A)	B1	NT
X611	HV Lisboa	Sick	M	5A	UD	AMP, CTX, CAZ, ATM, CHL, CIP, CN, TET	CTX-M-1	*tet*(A)	D	NT
X616	HV Lisboa	Sick	M	3M	UD	AMP, AUG, CTX, CAZ, TET	CTX-M-1	ND	B1	NT
X617	HV Lisboa	Sick	F	3A	UD	AMP, CTX, CAZ, TET	CTX-M-1	*tet*(A)	B1	NT
C10265	CV Bragança	Sick	M	1A	UD	AMP, CTX, CAZ, S	CTX-M-1	ND	A	NT
X555	HD	Healthy	M	6A	Pastor alemão	AMP, CTX, CAZ, ATM, CIP, SXT, TOB, CN, TET	CTX-M-55	*tet*(B), *int*1	A	ST617
X568	HD	Healthy	M	7A	UD	AMP, AUG, CTX, CAZ, ATM, CHL, CIP, SXT, TET	CTX-M-55	*tet*(A), *int*1	B1	NT
C10149	HVTM	Sick	F	1,5A	UD	AMP, CTX, CAZ, CHL, TOB, CN, S, TET	CTX-M-55, TEM	*tet*(A)	A	NT
X573	HD	Healthy	M	1A	UD	AMP, AUG, CTX, CAZ, ATM, CHL, SXT, S, TET	CTX-M-32	*int*1	A	ST5766
X561	Kennel	Healthy	M	2A	Gado transm.	AMP, AUG, CTX, ATM, CHL, CIP, SXT, TOB, CN, S, TET	CTX-M-32	*tet*(A), *int*1	B1	ST3078
X571	HD	Healthy	M	1A	UD	AMP, AUG, CTX, CAZ, CHL, SXT, S, TET	CTX-M-32, TEM	*tet*(B), *int*1	B1	NT
X572	HD	Healthy	F	1A	UD	AMP, AUG, CTX, CAZ, S, TET	CTX-M-14	*tet*(B)	A	ST542
X574	CVTransm	Sick	M	4A	UD	AMP, CTX, CHL, SXT, TOB, CN, S, TET	CTX-M-14	ND	A	NT
X565	HD	Healthy	M	6A	UD	AMP, CTX, ATM, CIP, SXT, TOB, CN, S, TET	CTX-M-variant	ND	A	NT
C10147	HVLisboa	Sick	M	7A	UD	AMP, CTX, CHL, SXT, CN, S, TET	TEM-1	*tet*(A), *int*1	B2	NT
X587	HVTM	Sick	M	2A	Bulldog Francês	AMP, CTX, ATM, CHL, SXT, CN, S, TET	No *bla* genes	*int*1	A	NT

[a] HD- healthy dogs from their owners; HVTM- *Hospital Veterinário de Trás os Montes* (Vila Real); Kennel-healthy dogs from the kennel (Vila Real); CV Transm- Clínica Veterinária Transmonvete (Vila Real, Portugal); HV Lisboa- *Hospital Veterinário de São Bento (Lisboa)*; CV Vouga- Clínica Veterinária do Vouga (Sever do Vouga, Portugal); CV Bragança- Clínica Veterinária de Macedo de Cavaleiros (Bragança, Portugal); CV VR- Clínica Veterinária dos Quinchosos (Vila Real, Portugal); [b] female; M-male; [c] A- years; M- months; [d] UD- undetermined dog breed; [e] AMP, ampicillin; AUG, amoxicillin–clavulanic acid; FOX, cefoxitin; CTX, cefotaxime; CAZ, ceftazidime; ATM, aztreonam; CHL, chloramphenicol; CIP, ciprofloxacin; TOB, tobramycin; AK, amikacin; CN, gentamicin; SXT, trimethoprim–sulfamethoxazole; S, streptomycin; TET, tetracycline; [f] ND: not detected; [g] Phylogroups; [h] MLST-Multilocus Sequence Typing; NT: not tested.

Table 2. Phenotypic and molecular features of ESBL-negative *E. coli* isolates recovered from healthy and sick dogs in Portugal.

Isolate Number	Origin [a]	Gender [b]	Age [c]	Breed [d]	Antimicrobial Resistance Phenotype [e]	Resistance Genotype	Other Resistance Genes [f]	PG [g]	MLST [h]
X551	HD	F	24M	Golden Retriever	AMP, CTX	CMY-2	ND	D	New ST *
X567	CV Vouga	F	8A	UD	AMP, AUG, FOX, CTX, CAZ, CIP, S, TET	CMY-2, TEM	*tet*(A)	D	ST115
X549	HVTM	F	6A	Leão Rodesea	AMP, AUG, CTX	ND	ND	D	NT
C10266	HV Lisboa	F	6A	UD	AMP, AUG, FOX, CTX, CAZ, ATM, NA, CIP, SXT, S, TET	ND	*tet*(B)	A	NT

[a] HD- healthy dogs from their owners; HVTM- *Hospital Veterinário de Trás os Montes* (Vila Real); Kennel- healthy dogs from kennel (Vila Real); CV Transm- Clínica Veterinária Transmonvete (Vila Real, Portugal); HV Lisboa- *Hospital Veterinário de São Bento (Lisboa)*; CV Vouga- Clínica Veterinária do Vouga (Sever do Vouga, Portugal); CV Bragança- Clínica Veterinária de Macedo de Cavaleiros (Bragança, Portugal); CV VR- Clínica Veterinária dos Quinchosos (Vila Real, Portugal); [b] F-female; M-male; [c] A- years; M- months; [d] UD- undetermined dog breed; [e] AMP, ampicillin; AUG, amoxicillin–clavulanic acid; FOX, cefoxitin; CTX, cefotaxime; CAZ, ceftazidime; ATM, aztreonam; NA, nalidixic acid; CIP, ciprofloxacin; SXT, trimethoprim–sulfamethoxazole; S, streptomycin; TET, tetracycline; IMP, imipenem; ETP, ertapenem. [f] ND: not detected; [g] Phylogroups; [h] MLST-Multilocus Sequence Typing; NT: not tested. * New ST allelic combination: *fum*C (26), *adk* (4), *pur*A (5), *icd* (25), *gyr*B (2), *rec*A (2), and *mdh* (5).

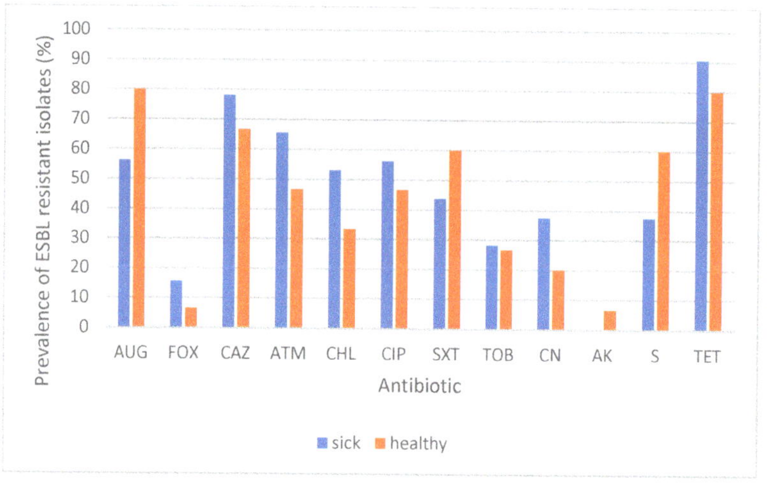

Figure 1. Prevalence of antibiotic-resistance among ESBL-producing *E. coli* isolates in sick and healthy dogs. Antibiotics tested: AUG, amoxicillin-clavulanic acid; FOX, cefoxitin; CAZ, ceftazidime; ATM, aztreonam; CHL, chloramphenicol; CIP, ciprofloxacin; SXT, trimethoprim-sulfamethoxazole; TOB, tobramycin; CN, gentamicin; AK, amikacin; S, streptomycin; TET, tetracycline. No significant association was detected between antibiotic resistance and type of animal (sick or healthy) ($p > 0.05$).

Different variants of bla_{CTX-M} genes were detected among 45 of these 47 ESBL-producing isolates (95.4%): $bla_{CTX-M-15}$ ($n = 26$ isolates), $bla_{CTX-M-1}$ ($n = 10$), $bla_{CTX-M-32}$ ($n = 3$), $bla_{CTX-M-55}$ ($n = 3$), $bla_{CTX-M-14}$ ($n = 2$), and bla_{CTX-M} ($n = 1$, no variant determined) (Table 1). Figure 2 shows the distribution of the ESBL variants depending on the origin of the isolates; no statistical difference could be established between the origin of the strain (healthy or sick dog) and the ESBL type ($p > 0.05$) (Figure 2), except for CTX-M-32, in which this relation was present (it was detected just in healthy dogs).

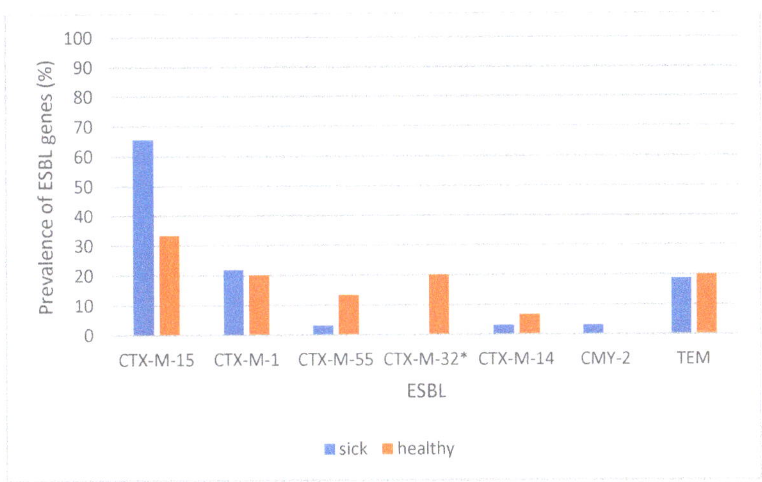

Figure 2. Distribution of ESBL-encoding genes from *E. coli* isolates in sick and healthy dogs. Gene encoding β-lactamases with $p < 0.05$ is indicated with (*).

Two of the four CTX^R and ESBL-negative isolates were CMY-2-producers (qAmpC type), and they were recovered from a healthy and a sick dog (one each) (Table 2). We could not detect the mechanisms of CTX^R in the two remaining ESBL-negative isolates. None of the CTX^R *E. coli* isolates carried the *mcr*-1 gene (related to colistin resistance).

Moreover, other β-lactamases genes such as bla_{VEB}, bla_{NDM}, bla_{OXA-48}, and bla_{VIM} were tested by PCR/sequencing but all isolates were revealed to be negative. Furthermore, the $stx_{1,2}$ genes related to Shiga toxin-producing *E. coli* (STEC) were not detected among our isolates.

The ESBL-positive isolates were ascribed to phylogenetic groups B_1 (n = 21 isolates), A (n = 14), D (n = 7), and B_2 (n = 5, two of them CTX-M-15-producers, typed as ST131) (Table 1). Furthermore, the four ESBL-negative isolates belonged to phylogroups D (n = 3, including the two CMY-2 producers) and A (n = 1) (Table 2).

MLST analysis, which was performed in thirteen representative *E. coli* isolates (based on the antimicrobial-resistance phenotype), revealed ten different lineages (ST/phylogenetic-group/β-lactamase): ST131/B2/CTX-M-15 (n = 2, from sick dogs, one from Lisbon and another from Bragança hospitals), ST617/A/CTX-M-55 (n = 1, from a healthy dog), ST3078/B_1/CTX-M-32 (n = 1, from a healthy dog from the kennel), ST57/D/CTX-M-1 (n = 1, from a sick dog from Vouga clinic), ST12/B_2/CTX-M-15 (n = 2 sick dogs, one from Vila Real and another from Lisbon), ST6448/B_1/CTX-M-15 (n = 2 sick dogs, one of them CMY-2 positive and both from Lisbon), ST542/A/CTX-M-14 (n = 1, from a healthy dog), ST5766/A/CTX-M-32 (n = 1, from a healthy dog), and ST115/D/CMY-2 (n = 1, from a sick dog from Vouga clinic); moreover, one CMY-2-producing *E. coli* isolate of phylogroup D obtained in a healthy dog, presented a new combination of alleles (*fum*C (26), *adk* (4), *pur*A (5), *icd* (25), *gyr*B (2), *rec*A (2) and *mdh* (5)), rendering a new ST (Table 1).

4. Discussion

Regarding the Portuguese situation, the prevalence of ESBL-producing *E. coli* isolates in healthy dogs obtained in this work is similar to previous studies performed in dogs and cats [12,22,23] in the South and the North of Portugal. Worldwide, this prevalence was lower than the ones obtained with faecal samples of healthy dogs in Germany, Brazil, or China (24–29%) [15,37,38], but it is similar to the results of previous studies performed in Tunisia and France (12.7–17%) [11,39]. These differences could be explained by differences in the epidemiology of ESBL genes among different countries, considering the year in which the studies were performed, but we cannot discard methodological effects in the different studies.

Five types of CTX-M ESBLs were detected, indicating a high diversity of CTX-M genes (mainly $bla_{CTX-M-15}$ gene) among the CTX^R *E. coli* isolates; these results are in accordance with a previous study done in Portugal on healthy dogs [12]. This $bla_{CTX-M-15}$ gene was also the most frequently detected in *E. coli* isolated from dogs in different countries [3,15,40]. The CTX-M-1- and CTX-M-15-encoding genes were also detected among *E. coli* canine isolates in Italy [41] and Denmark [13], which are in agreement with our data. The same variants of CTX-M genes were observed in a recent study conducted on healthy humans in Spain [42]. Moreover, during the last few years, new variants are becoming more common, in particular CTX-M-55 [3], especially from companion animals in Asian countries [43].

In the past, the $bla_{CTX-M-15}$ gene was mainly associated with strains of human origin while $bla_{CTX-M-1}$ was the major CTX-M sub-type among livestock and companion animal isolates in Europe [15,41]. Actually, this close correspondence is no longer so obvious, and our results confirm these data. A further study should be implemented to determine the ESBL gene in the two uncharacterized ESBL-producing isolates.

In this study, the CMY-2 gene was the qAmpC β-lactamase type found among two CTX^R-ESBL-negative isolates and one ESBL-producing isolate, and it has been previously reported among *E. coli* strains from healthy and sick pets worldwide [20,23,39,44]. The detection of *tet*A and/or *tet*B genes in most of our tetracycline-resistant isolates seem to be similar to the results obtained by Costa et al. [45] from dogs, in Northern Portugal.

In this work, the most common phylogenetic groups among our isolates were B$_1$ and A, these being the phylogroups more associated with commensal *E. coli* both in humans and in dogs, as well as in other animals [11,13]. On the other hand, isolates belonging to phylogroup B$_2$ and D are more likely to be recovered from extra-intestinal infections of companion animals [4]. An interesting study related to 78 dogs that visited a veterinary hospital in Northern Portugal (either for a normal checkout or in case of disease) revealed the prevalence of *E. coli* isolates of groups A ($n = 19$), D ($n = 9$), and B$_1$ ($n = 7$) [46], similar to our observation. So, the carriage of ESBL/qAmpC producing *E. coli* of these phylogroups in the gastrointestinal tract suggests a potential reservoir of MDR ESBL-producing bacteria in dogs.

Regarding the MLST results, the pandemic virulent *E. coli* ST131-B2 clone was detected among two isolates of sick dogs tested in this study. It is important to note that this clone was widely detected in pets [47,48], including in sick dogs in Portugal [17,49].

On the other hand, we detected one *E. coli* strain, ST57/D/CTX-M-1, that was recently detected in Portugal (associated with CMY-2 gene) in a dog with a UTI from a Lisbon hospital [17]. Similarly, the same lineage was identified in a faecal isolate from a healthy dog in Mexico, characterized as CMY-2/ST57/D) [50].

The frequency of the ST6448 lineage, which was observed in two sick dogs in this study, is considered an infrequent clone in humans and companion animals. This lineage was also found among a vulture faecal sample from Canary Islands [51]. To our knowledge, there is only one previous report related to the detection of this clone in humans, which was recently reported in healthy children from Sweden [52].

Additionally, our data indicate the presence of *E. coli* ST12/B2/CTX-M-15, which should be considered an agent of high clinical relevance for humans and animals. Furthermore, the ST12 lineage (associated with CMY-2) was identified in healthy dogs from Spain [6], Brazil [2], and France [11]. Furthermore, this lineage was found among isolates from children with a febrile UTI in France [53] and in healthy humans in Spain [42]. These findings highlight the dissemination of ST12 lineage and its presence in animal and human' isolates.

To our knowledge, the ST617 lineage (clonal complex ST10) was identified for the first time in pets from Portugal in this study. CTX-M-15-producing *E. coli* isolates of sequence type ST617/phylogroup A have been reported in sick dogs in France [40] and in hospitalized patients in Tunisia [54,55]. Similarly, Rocha-Gracia et al. [50] identified the same lineage among a faecal isolate from healthy dogs in Mexico (ST617/A/CTX-M-15). According to a recent study, Gauthier, et al. [56] found this lineage in four isolates from dogs in France harbouring carbapenemase genes. Furthermore, this clone was widely disseminated.

The ST542 lineage detected in one of the healthy dogs is not commonly reported; however, this clone was found in a farmworker from Germany [57] and in a pig in Australia [58]. On the other hand, an ST115/CMY-2 isolate (found in a sick dog from the Vouga clinic) was previously reported among chickens and human patients in Germany [47].

We also detected a ST5766/A/CTX-M-32 isolate in a healthy dog; this clone is unusual, and it was previously reported in broilers' osteomyelitis in Brazil [59]. To our knowledge, this is the first report of the ST5766 clone among pets, and the first detection in Europe. In this study, we also found an *E. coli* isolate, ST3078/B1/CTX-M-32, recovered from a healthy dog from a kennel. To our knowledge, the only unique previous study related to the ST3078 lineage was found in wastewater in Eastern France [60]. This suggests that the environment likely plays a role in the spread of ESBL-producing *E. coli* isolates in the community, associated with a One Health approach (human-animals-environment). Importantly, a new combination of alleles was found in an isolate of a healthy dog, rendering a new ST.

The use of β-lactams in the clinical practice of veterinary medicine may be considered one of the reasons for the high incidence of ESBL-producers worldwide. Thus, pets can be a significant source of ESBL/qAmpC-producing *E. coli* isolates. Considering the prevalence of ESBLs (notably the large reservoir in dogs of *E. coli* isolates with genes encoding CTX-M-15 and CTX-M-1, or CMY-2 β-lactamases), there is a serious and plausible risk of future acquisition of these resistant genes by their owners.

5. Conclusions

Antimicrobial resistance can make infections difficult to treat, which represents a global public health problem, due to the negative consequences for human health. This study shows that healthy and sick dogs are frequent carriers of faecal ESBL-producing *E. coli* strains, harbouring different variants of bla_{CTX-M} genes (mostly $bla_{CTX-M-15}$ and $bla_{CTX-M-1}$), and presenting a high genetic MLST diversity (including the ST131/B2 lineage). Although at a lower rate, the bla_{CMY-2} gene was also found. This fact suggests the implication of mobile genetic elements in the dissemination of this relevant mechanism of resistance. This underlies the complexity of the antimicrobial resistance of bacteria occurring in dogs and the possible interspecies transmission between humans, domestic animals, and into the environment, important knowledge given the One-Health approach.

Supplementary Materials: The following are available online at https://www.mdpi.com/article/10.3390/antibiotics10081013/s1, Figure S1: Geographic location of the different areas where the faecal samples from sick dogs were collected in Portugal. Table S1: Primers sequences and PCR conditions used for genes encoding antibiotic resistance in *E. coli*.

Author Contributions: I.C.: conceptualization, sampling, methodology, investigation, resources, data curation, writing and rewriting; R.C.(Rosa Capita): sampling, methodology; C.M.: sampling, methodology; S.M.-Á.: investigation, data curation; N.S.C.: investigation, data curation, writing—review and editing; P.P. (Paulo Pimenta): sampling, methodology; A.R.P.: sampling, methodology; S.R.: sampling, methodology; M.S.: investigation, data curation; Â.M.: statistical analyses; J.F.: sampling, methodology; F.R.: review and editing; R.C. (Rita Cunha): validation, funding acquisition; C.A.-C.: validation, funding acquisition; M.d.L.N.E.D.: review and editing; G.I.: conceptualization, methodology, validation, resources, data curation, writing—review and editing, visualization, supervision, project administration, funding acquisition; C.T.: conceptualization, methodology, validation, resources, data curation, writing—review and editing, visualization, supervision, project administration, funding acquisition; P.P. (Patrícia Poeta): conceptualization, methodology, validation, resources, data curation, writing—review and editing, visualization, supervision, project administration, funding acquisition. All authors have read and agreed to the published version of the manuscript.

Funding: I.C. gratefully acknowledges the financial support of "Fundação para a Ciência e Tecnologia" (FCT—Portugal) related to PhD grant, through the reference SFRH/BD/133266/2017 (Medicina Clínica e Ciências da Saúde), as well as MCTES (Ministério da Ciência, Tecnologia e Ensino Superior) and European Union (EU), with reference to Fundo Social Europeu (FSE). The experimental work carried out in the University of La Rioja (Spain) was financed by the project SAF2016-76571-R from the Agencia Estatal de Investigación (AEI) of Spain and FEDER of EU. N.S.C. was awarded a grant for the year 2018, from the Algerian Ministry of Higher Education and Scientific Research (The PNE Program), under the direction of Carmen Torres. This work was supported by the *Ministerio de Ciencia, Innovación y Universidades* (Spain; grant number RTI2018-098267-R-C33), the *Junta de Castilla y León* (*Consejería de Educación*, Spain; grant number LE018P20) and the Associate Laboratory for Green Chemistry—LAQV which is financed by national funds from FCT/MCTES (UIDB/50006/2020 and UIDP/50006/2020).

Institutional Review Board Statement: Not applicable.

Informed Consent Statement: Not applicable.

Data Availability Statement: The data presented in this study are available in Supplementary Material.

Acknowledgments: The authors would like to thank Carla Oliveira from Transmonvete Veterinary Clinic, and also Anabela Pais from Quinchosos Veterinary center who helped us with the collection of faecal samples of pets in Vila Real (Portugal). In the same line, we also appreciate the collaboration with the local kennel located inVila Real.

Conflicts of Interest: The authors declare no conflict of interest. Preliminary data was presented as a poster entitled "Diversity of CTX-M β-lactamases in *Escherichia coli* isolated from dogs in Portugal", in ECCMID 2019 Congress (Amsterdam), and as part of an awarded oral presentation named "High frequency of ESBL- *E. coli* producers in pets in Portugal with detection of ST131 clone carrying different variants of CTX-M genes", in IC2AR 2019 (Portugal).

References

1. Chen, Y.; Liu, Z.; Zhang, Y.; Zhang, Z.; Lei, L.; Xia, Z. Increasing Prevalence of ESBL-Producing Multidrug Resistance *Escherichia coli* From Diseased Pets in Beijing, China From 2012 to 2017. *Front. Microbiol.* **2019**, *10*, 2852. [CrossRef]

2. de Melo, L.C.; Oresco, C.; Leigue, L.; Netto, H.M.; Melville, P.A.; Benites, N.R.; Saras, E.; Haenni, M.; Lincopan, N.; Madec, J.-Y. Prevalence and molecular features of ESBL/pAmpC-producing *Enterobacteriaceae* in healthy and diseased companion animals in Brazil. *Vet. Microbiol.* **2018**, *221*, 59–66. [CrossRef]

3. Bortolami, A.; Zendri, F.; Maciuca, E.I.; Wattret, A.; Ellis, C.; Schmidt, V.; Pinchbeck, G.; Timofte, D. Diversity, Virulence, and Clinical Significance of Extended-Spectrum β-Lactamase- and pAmpC-Producing *Escherichia coli* From Companion Animals. *Front. Microbiol.* **2019**, *10*, 1260. [CrossRef] [PubMed]

4. Bourne, J.A.; Chong, W.L.; Gordon, D.M. Genetic structure, antimicrobial resistance and frequency of human associated *Escherichia coli* sequence types among faecal isolates from healthy dogs and cats living in Canberra, Australia. *PLoS ONE* **2019**, *14*, e0212867. [CrossRef] [PubMed]

5. WHO. Global Priority List of Antibiotic-Resistant Bacteria to Guide Research, Discovery, and Development of New Antibiot-ics. 2017. Available online: https://www.who.int/medicines/publications/WHO-PPL-Short_Summary_25Feb-ET_NM_WHO.pdf (accessed on 15 January 2021).

6. Dupouy, V.; Abdelli, M.; Moyano, G.; Arpaillange, N.; Bibbal, D.; Cadiergues, M.-C.; Lopez-Pulin, D.; Sayah-Jeanne, S.; De Gunzburg, J.; Saint-Lu, N.; et al. Prevalence of Beta-Lactam and Quinolone/Fluoroquinolone Resistance in *Enterobacteriaceae* From Dogs in France and Spain—Characterization of ESBL/pAmpC Isolates, Genes, and Conjugative Plasmids. *Front. Vet. Sci.* **2019**, *6*, 279. [CrossRef] [PubMed]

7. Carvalho, I.; Silva, N.; Carrola, J.; Silva, V.; Currie, C.; Igrejas, G.; Poeta, P. Immunity-Acquired Resistance: Evolution of Antimicrobial Resistance Among Extended-Spectrum β-Lactamases and Carbapenemases in Klebsiella pneumoniae and *Escherichia coli*. In *Antibiotic Drug Resistance Chapter 11*; John Wiley & Sons: Hoboken, NJ, USA, 2019; pp. 239–259. [CrossRef]

8. Salgado-Caxito, M.; Benavides, J.A.; Adell, A.D.; Paes, A.C.; Moreno-Switt, A.I. Global prevalence and molecular characterization of extended-spectrum β-lactamase producing-*Escherichia coli* in dogs and cats—A scoping review and meta-analysis. *One Health* **2021**, *12*, 100236. [CrossRef] [PubMed]

9. Benavides, J.A.; Salgado-Caxito, M.; Opazo-Capurro, A.; González Muñoz, P.; Piñeiro, A.; Otto Medina, M.; Rivas, L.; Munita, J.; Millán, J. ESBL-Producing *Escherichia coli* Carrying CTX-M Genes Circulating among Livestock, Dogs, and Wild Mammals in Small-Scale Farms of Central Chile. *Antibiotics* **2021**, *10*, 510. [CrossRef] [PubMed]

10. Li, J.; Bi, Z.; Ma, S.; Chen, B.; Cai, C.; He, J.; Schwarz, S.; Sun, C.; Zhou, Y.; Yin, J.; et al. Inter-host Transmission of Carbapenemase-Producing *Escherichia coli* among Humans and Backyard Animals. *Environ. Health Perspect.* **2019**, *127*, 107009. [CrossRef]

11. Haenni, M.; Saras, E.; Métayer, V.; Médaille, C.; Madec, J.-Y. High Prevalence of *bla*$_{CTX-M-1}$/IncI1/ST3 and *bla*$_{CMY-2}$/IncI1/ST2 Plasmids in Healthy Urban Dogs in France. *Antimicrob. Agents Chemother.* **2014**, *58*, 5358–5362. [CrossRef]

12. Belas, A.; Salazar, A.S.; Gama, L.; Couto, N.; Pomba, C. Risk factors for faecal colonisation with *Escherichia coli* producing extended-spectrum and plasmid-mediated AmpC β-lactamases in dogs. *Vet. Rec.* **2014**, *175*, 202. [CrossRef]

13. Damborg, P.; Morsing, M.K.; Petersen, T.; Bortolaia, V.; Guardabassi, L. CTX-M-1 and CTX-M-15-producing *Escherichia coli* in dog faeces from public gardens. *Acta Vet. Scand.* **2015**, *57*, 83. [CrossRef]

14. Gómez-Sanz, E.; Torres, C.; Ceballos, S.; Lozano, C.; Zarazaga, M. Clonal Dynamics of Nasal *Staphylococcus aureus* and *Staphylococcus pseudintermedius* in Dog-Owning Household Members. Detection of MSSA ST398. *PLoS ONE* **2013**, *8*, e69337. [CrossRef] [PubMed]

15. Boehmer, T.; Vogler, A.J.; Thomas, A.; Sauer, S.; Hergenroether, M.; Straubinger, R.K.; Birdsell, D.; Keim, P.; Sahl, J.W.; Williamson, C.H.D.; et al. Phenotypic characterization and whole genome analysis of extended-spectrum beta-lactamase-producing bacteria isolated from dogs in Germany. *PLoS ONE* **2018**, *13*, e0206252. [CrossRef]

16. Kaspar, U.; Von Lützau, A.; Schlattmann, A.; Roesler, U.; Köck, R.; Becker, K. Zoonotic multidrug-resistant microorganisms among small companion animals in Germany. *PLoS ONE* **2018**, *13*, e0208364. [CrossRef] [PubMed]

17. Marques, C.; Belas, A.; Franco, A.; Aboim, C.; Gama, L.; Pomba, C. Increase in antimicrobial resistance and emergence of major international high-risk clonal lineages in dogs and cats with urinary tract infection: 16 year retrospective study. *J. Antimicrob. Chemother.* **2017**, *73*, 377–384. [CrossRef]

18. Zogg, A.L.; Simmen, S.; Zurfluh, K.; Stephan, R.; Schmitt, S.N.; Nüesch-Inderbinen, M. High Prevalence of Extended-Spectrum β-Lactamase Producing *Enterobacteriaceae* Among Clinical Isolates from Cats and Dogs Admitted to a Veterinary Hospital in Switzerland. *Front. Vet. Sci.* **2018**, *5*, 62. [CrossRef] [PubMed]

19. Nigg, A.; Brilhante, M.; Dazio, V.; Clément, M.; Collaud, A.; Brawand, S.G.; Willi, B.; Endimiani, A.; Schuller, S.; Perreten, V. Shedding of OXA-181 carbapenemase-producing *Escherichia coli* from companion animals after hospitalisation in Switzerland: An outbreak in 2018. *Eurosurveillance* **2019**, *24*, 1900071. [CrossRef] [PubMed]

20. Hordijk, J.; Schoormans, A.; Kwakernaak, M.; Duim, B.; Broens, E.; Dierikx, C.; Mevius, D.; Wagenaar, J.A. High prevalence of fecal carriage of extended spectrum β-lactamase/AmpC-producing *Enterobacteriaceae* in cats and dogs. *Front. Microbiol.* **2013**, *4*, 242. [CrossRef]

21. Brilhante, M.; Menezes, J.; Belas, A.; Feudi, C.; Schwarz, S.; Pomba, C.; Perreten, V. OXA-181-Producing Extraintestinal Pathogenic *Escherichia coli* Sequence Type 410 Isolated from a Dog in Portugal. *Antimicrob. Agents Chemother.* **2020**, *64*, e02298-19. [CrossRef]

22. Costa, D.; Poeta, P.; Briñas, L.; Sáenz, Y.; Rodrigues, J.; Torres, C. Detection of CTX-M-1 and TEM-52 β-lactamases in *Escherichia coli* strains from healthy pets in Portugal. *J. Antimicrob. Chemother.* **2004**, *54*, 960–961. [CrossRef]

23. Carvalho, I.; Safia Chenouf, N.; Cunha, R.; Martins, C.; Pimenta, P.; Pereira, A.R.; Martínez-Álvarez, S.; Ramos, S.; Silva, V.; Igrejas, G.; et al. Antimicrobial Resistance Genes and Diversity of Clones among ESBL—And Acquired AmpC-Producing *Escherichia coli* Isolated from Fecal Samples of Healthy and Sick Cats in Portugal. *Antibiotics* **2021**, *10*, 262. [CrossRef] [PubMed]

24. CLSI. Clinical and Laboratory Standards Institute. In *Performed Standards for Antimicrobial Susceptibility Testing*, 29th ed.; CLSI Supplement: Wayne, PA, USA, 2019.

25. Holmes, D.; Quigley, M. A rapid boiling method for the preparation of bacterial plasmids. *Anal. Biochem.* **1981**, *114*, 193–197. [CrossRef]

26. Ruiz, E.; Sáenz, Y.; Zarazaga, M.; Rocha-Gracia, R.; Martínez, L.M.; Arlet, G.; Torres, C. qnr, aac(6′)-Ib-cr and qepA genes in *Escherichia coli* and *Klebsiella* spp.: Genetic environments and plasmid and chromosomal location. *J. Antimicrob. Chemother.* **2012**, *67*, 886–897. [CrossRef] [PubMed]

27. Zong, Z.; Partridge, S.R.; Thomas, L.; Iredell, J.R. Dominance of *bla*$_{CTX-M}$ within an Australian Extended-Spectrum β-Lactamase Gene Pool. *Antimicrob. Agents Chemother.* **2008**, *52*, 4198–4202. [CrossRef]

28. Pitout, J.D.; Thomson, K.S.; Hanson, N.D.; Ehrhardt, A.F.; Moland, E.S.; Sanders, C.C. β-Lactamases responsible for resistance to expanded-spectrum cephalosporins in *Klebsiella pneumoniae*, *Escherichia coli*, and *Proteus mirabilis* isolates recovered in South Africa. *Antimicrob. Agents Chemother.* **1998**, *42*, 1350–1354. [CrossRef]

29. Porres-Osante, N.; Azcona-Gutiérrez, J.M.; Rojo-Bezares, B.; Undabeitia, E.; Torres, C.; Sáenz, Y. Emergence of a multiresistant KPC-3 and VIM-1 carbapenemase-producing *Escherichia coli* strain in Spain. *J. Antimicrob. Chemother.* **2014**, *69*, 1792–1795. [CrossRef]

30. Vinué, L.; Sáenz, Y.; Somalo, S.; Escudero, E.; Moreno, M.A.; Ruiz-Larrea, F.; Torres, C. Prevalence and diversity of integrons and associated resistance genes in faecal *Escherichia coli* isolates of healthy humans in Spain. *J. Antimicrob. Chemother.* **2008**, *62*, 934–937. [CrossRef]

31. Hassen, B.; Abbassi, M.S.; Ruiz-Ripa, L.; Mama, O.M.; Hassen, A.; Torres, C.; Hammami, S. High prevalence of mcr-1 encoding colistin resistance and first identification of blaCTX-M-55 in ESBL/CMY-2-producing *Escherichia coli* isolated from chicken faeces and retail meat in Tunisia. *Int. J. Food Microbiol.* **2019**, *318*, 108478. [CrossRef]

32. Ellington, M.J.; Kistler, J.J.; Livermore, D.M.; Woodford, N. Multiplex PCR for rapid detection of genes encoding acquired metallo-lactamases. *J. Antimicrob. Chemother.* **2006**, *59*, 321–322. [CrossRef]

33. Vidal, R.; Vidal, M.; Lagos, R.; Levine, M.; Prado, V. Multiplex PCR for Diagnosis of Enteric Infections Associated with Diarrheagenic *Escherichia coli*. *J. Clin. Microbiol.* **2004**, *42*, 1787–1789. [CrossRef]

34. PubMLST. *Escherichia coli* (Achtman) MLST Locus/Sequence Definitions Database. Available online: https://pubmlst.org/bigsdb?db=pubmlst_ecoli_achtman_seqdef (accessed on 15 January 2021).
35. Clermont, O.; Bonacorsi, S.; Bingen, E. Rapid and Simple Determination of the *Escherichia coli* Phylogenetic Group. *Appl. Environ. Microbiol.* **2000**, *66*, 4555–4558. [CrossRef] [PubMed]
36. Jara, D.; Bello-Toledo, H.; Domínguez, M.; Cigarroa, C.; Fernández, P.; Vergara, L.; Quezada-Aguiluz, M.; Opazo-Capurro, A.; Lima, C.; González-Rocha, G. Antibiotic resistance in bacterial isolates from freshwater samples in Fildes Peninsula, King George Island, Antarctica. *Sci. Rep.* **2020**, *10*, 1–8. [CrossRef] [PubMed]
37. Carvalho, A.; Barbosa, A.; Arais, L.; Ribeiro, P.; Carneiro, V.; Cerqueira, A. Resistance patterns, ESBL genes, and genetic relatedness of *Escherichia coli* from dogs and owners. *Braz. J. Microbiol.* **2016**, *47*, 150–158. [CrossRef] [PubMed]
38. Sun, Y.; Zeng, Z.; Chen, S.; Ma, J.; He, L.; Liu, Y.; Deng, Y.; Lei, T.; Zhao, J.; Liu, J.-H. High prevalence of bla_{CTX-M} extended-spectrum β-lactamase genes in *Escherichia coli* isolates from pets and emergence of CTX-M-64 in China. *Clin. Microbiol. Infect.* **2010**, *16*, 1475–1481. [CrossRef] [PubMed]
39. Ben Sallem, R.; Gharsa, H.; BEN Slama, K.; Rojo-Bezares, B.; Estepa, V.; Porres-Osante, N.; Jouini, A.; Klibi, N.; Sáenz, Y.; Boudabous, A.; et al. First Detection of CTX-M-1, CMY-2, and QnrB19 Resistance Mechanisms in Fecal *Escherichia coli* Isolates from Healthy Pets in Tunisia. *Vector Borne Zoonotic Dis.* **2013**, *13*, 98–102. [CrossRef] [PubMed]
40. Dahmen, S.; Haenni, M.; Châtre, P.; Madec, J.-Y. Characterization of bla_{CTX-M} IncFII plasmids and clones of *Escherichia coli* from pets in France. *J. Antimicrob. Chemother.* **2013**, *68*, 2797–2801. [CrossRef]
41. Iseppi, R.; Di Cerbo, A.; Messi, P.; Sabia, C. Antibiotic Resistance and Virulence Traits in Vancomycin-Resistant Enterococci (VRE) and Extended-Spectrum β-Lactamase/AmpC-producing (ESBL/AmpC) *Enterobacteriaceae* from Humans and Pets. *Antibiotics* **2020**, *9*, 152. [CrossRef]
42. Rodriguez-Navarro, J.; Miro, E.; Brown-Jaque, M.; Hurtado, J.C.; Moreno, A.; Muniesa, M.; Gonzalez-Lopez, J.J.; Vila, J.; Espinal, P.; Navarro, F. Diversity of plasmids in *Escherichia coli* and *Klebsiella pneumoniae*: A comparison of commensal and clinical isolates. *Antimicrob. Agents Chemother.* **2020**, *64*. [CrossRef]
43. Kawamura, K.; Sugawara, T.; Matsuo, N.; Hayashi, K.; Norizuki, C.; Tamai, K.; Kondo, T.; Arakawa, Y. Spread of CTX-Type Extended-Spectrum β-Lactamase-Producing *Escherichia coli* Isolates of Epidemic Clone B2-O25-ST131 Among Dogs and Cats in Japan. *Microb. Drug Resist.* **2017**, *23*, 1059–1066. [CrossRef]
44. Aslantaş, O.; Yilmaz, E. Prevalence and molecular characterization of extended-spectrum β-lactamase (ESBL) and plasmidic AmpC β-lactamase (pAmpC) producing *Escherichia coli* in dogs. *J. Vet. Med Sci.* **2017**, *79*, 1024–1030. [CrossRef]
45. Costa, D.; Poeta, P.; Sáenz, Y.; Coelho, A.C.; Matos, M.; Vinue, L.; Rodrigues, J.; Torres, C. Prevalence of antimicrobial resistance and resistance genes in faecal *Escherichia coli* isolates recovered from healthy pets. *Vet. Microbiol.* **2008**, *127*, 97–105. [CrossRef]
46. Meireles, D.; Leite-Martins, L.; Bessa, L.; Cunha, S.; Fernandes, R.; De Matos, A.; Manaia, C.; Da Costa, P.M. Molecular characterization of quinolone resistance mechanisms and extended-spectrum β-lactamase production in *Escherichia coli* isolated from dogs. *Comp. Immunol. Microbiol. Infect. Dis.* **2015**, *41*, 43–48. [CrossRef] [PubMed]
47. Pietsch, M.; Irrgang, A.; Roschanski, N.; Michael, G.B.; Hamprecht, A.; Rieber, H.; Käsbohrer, A.; Schwarz, S.; Rösler, U. Whole genome analyses of CMY-2-producing *Escherichia coli* isolates from humans, animals and food in Germany. *BMC Genom.* **2018**, *19*, 601. [CrossRef]
48. Liakopoulos, A.; Betts, J.; La Ragione, R.; van Essen-Zandbergen, A.; Ceccarelli, D.; Petinaki, E.; Koutinas, C.K.; Mevius, D.J. Occurrence and characterization of extended-spectrum cephalosporin-resistant *Enterobacteriaceae* in healthy household dogs in Greece. *J. Med. Microbiol.* **2018**, *67*, 931–935. [CrossRef]
49. Belas, A.; Marques, C.; Aboim, C.; Pomba, C. Emergence of *Escherichia coli* ST131 H30/H30-Rx subclones in companion animals. *J. Antimicrob. Chemother.* **2018**, *74*, 266–269. [CrossRef]
50. Rocha-Gracia, R.; Cortés-Cortés, G.; Lozano-Zarain, P.; Bello, F.; Martínez-Laguna, Y.; Torres, C. Faecal *Escherichia coli* isolates from healthy dogs harbour CTX-M-15 and CMY-2 β-lactamases. *Vet. J.* **2014**, *203*, 315–319. [CrossRef]
51. Carvalho, I.; Tejedor-Junco, M.T.; González-Martín, M.; Corbera, J.A.; Suárez-Pérez, A.; Silva, V.; Igrejas, G.; Torres, C.; Poeta, P. Molecular diversity of Extended-spectrum β-lactamase-producing *Escherichia coli* from vultures in Canary Islands. *Environ. Microbiol. Rep.* **2020**, *12*, 540–547. [CrossRef] [PubMed]
52. Kaarme, J.; Riedel, H.; Schaal, W.; Yin, H.; Nevéus, T.; Melhus, Å. Rapid Increase in Carriage Rates of *Enterobacteriaceae* Producing Extended-Spectrum β-Lactamases in Healthy Preschool Children, Sweden. *Emerg. Infect. Dis.* **2018**, *24*, 1874–1881. [CrossRef] [PubMed]
53. Birgy, A.; Madhi, F.; Jung, C.; Levy, C.; Cointe, A.; Bidet, P.; Hobson, C.A.; Bechet, S.; Sobral, E.; Vuthien, H.; et al. Diversity and trends in population structure of ESBL-producing *Enterobacteriaceae* in febrile urinary tract infections in children in France from 2014 to 2017. *J. Antimicrob. Chemother.* **2019**, *75*, 96–105. [CrossRef]
54. Mhaya, A.; Trabelsi, R.; Begu, D.; Aillerie, S.; M'zali, F.; Tounsi, S.; Gdoura, R.; Arpin, C. Emergence of B2-ST131-C2 and A-ST617 *Escherichia coli* clones producing both CTX-M-15 and CTX-M-27 and ST147 NDM-1 positive *Klebsiella pneumoniae* in the Tunisian community. *bioRxiv* **2019**, 713461. [CrossRef]
55. Ben Sallem, R.; Ben Slama, K.; Estepa, V.; Cheikhna, E.O.; Mohamed, A.M.; Chairat, S.; Ruiz-Larrea, F.; Boudabous, A.; Torres, C. Detection of CTX-M-15-producing *Escherichia coli* isolates of lineages ST410-A, ST617-A and ST354-D in faecal samples of hospitalized patients in a Mauritanian hospital. *J. Chemother.* **2014**, *27*, 114–116. [CrossRef]

56. Gauthier, L.; Dortet, L.; Cotellon, G.; Creton, E.; Cuzon, G.; Ponties, V.; Bonnin, R.A.; Naas, T. Diversity of Carbapenemase-Producing *Escherichia coli* Isolates in France in 2012–2013. *Antimicrob. Agents Chemother.* **2018**, *62*, e00266-18. [CrossRef] [PubMed]

57. Dahms, C.; Hübner, N.-O.; Kossow, A.; Mellmann, A.; Dittmann, K.; Kramer, A. Occurrence of ESBL-Producing *Escherichia coli* in Livestock and Farm Workers in Mecklenburg-Western Pomerania, Germany. *PLoS ONE* **2015**, *10*, e0143326. [CrossRef] [PubMed]

58. Reid, C.; Wyrsch, E.R.; Chowdhury, P.R.; Zingali, T.; Liu, M.; Darling, A.; Chapman, T.; Djordjevic, S.P. Porcine commensal *Escherichia coli*: A reservoir for class 1 integrons associated with IS26. *Microb. Genom.* **2017**, *3*, e000143. [CrossRef] [PubMed]

59. Braga, J.F.V.; Chanteloup, N.K.; Trotereau, A.; Baucheron, S.; Guabiraba, R.; Ecco, R.; Schouler, C. Diversity of *Escherichia coli* strains involved in vertebral osteomyelitis and arthritis in broilers in Brazil. *BMC Vet. Res.* **2016**, *12*, 1–12. [CrossRef] [PubMed]

60. Bréchet, C.; Plantin, J.; Thouverez, M.; Cholley, P.; Bertrand, X.; Hocquet, D. The wastewater network likely plays a role in the spread of ESBL-producing *Escherichia coli* in the community. In Proceedings of the 23rd European Society of Clinical Microbiology and Infectious Diseases (ESCMID), Berlin, Germany, 12 February 1983.

Article

Virulence Factors in *Staphylococcus* Associated with Small Ruminant Mastitis: Biofilm Production and Antimicrobial Resistance Genes

Nara Cavalcanti Andrade [1], Marta Laranjo [1], Mateus Matiuzzi Costa [2] and Maria Cristina Queiroga [1,3,*]

[1] MED–Mediterranean Institute for Agriculture, Environment and Development, Instituto de Investigação e Formação Avançada, Universidade de Évora, Pólo da Mitra, Ap. 94, 7006-554 Évora, Portugal; naraegabriel@hotmail.com (N.C.A.); mlaranjo@uevora.pt (M.L.)

[2] Federal University of the São Francisco Valley, BR 407 Highway, Nilo Coelho Irrigation Project, s/n C1, Petrolina 56300-000, PE, Brazil; mateus.costa@univasf.edu.br

[3] Departamento de Medicina Veterinária, Escola de Ciências e Tecnologia, Universidade de Évora, Pólo da Mitra, Ap. 94, 7006-554 Évora, Portugal

* Correspondence: crique@uevora.pt; Tel.: +351-266-740-800

Abstract: Small ruminant mastitis is a serious problem, mainly caused by *Staphylococcus* spp. Different virulence factors affect mastitis pathogenesis. The aim of this study was to investigate virulence factors genes for biofilm production and antimicrobial resistance to β-lactams and tetracyclines in 137 staphylococcal isolates from goats (86) and sheep (51). The presence of *coa*, *nuc*, *bap*, *icaA*, *icaD*, *blaZ*, *mecA*, *mecC*, *tetK*, and *tetM* genes was investigated. The *nuc* gene was detected in all *S. aureus* isolates and in some coagulase-negative staphylococci (CNS). None of the *S. aureus* isolates carried the *bap* gene, while 8 out of 18 CNS harbored this gene. The *icaA* gene was detected in *S. aureus* and *S. warneri*, while *icaD* only in *S. aureus*. None of the isolates carrying the *bap* gene harbored the *ica* genes. None of the biofilm-associated genes were detected in 14 isolates (six *S. aureus* and eight CNS). An association was found between *Staphylococcus* species and resistance to some antibiotics and between antimicrobial resistance and animal species. Nine penicillin-susceptible isolates exhibited the *blaZ* gene, questioning the reliability of susceptibility testing. Most *S. aureus* isolates were susceptible to tetracycline, and no cefazolin or gentamycin resistance was detected. These should replace other currently used antimicrobials.

Keywords: mastitis; staphylococci; virulence factors; genes; biofilm; antimicrobial resistance

Citation: Andrade, N.C.; Laranjo, M.; Costa, M.M.; Queiroga, M.C. Virulence Factors in *Staphylococcus* Associated with Small Ruminant Mastitis: Biofilm Production and Antimicrobial Resistance Genes. *Antibiotics* **2021**, *10*, 633. https://doi.org/10.3390/antibiotics10060633

Academic Editors: Elisabete Silva and Manuela Oliveira

Received: 26 April 2021
Accepted: 18 May 2021
Published: 25 May 2021

1. Introduction

Mastitis is the inflammation of the mammary gland, mainly due to intramammary infection (IMI). In small ruminants, this disease is considered a serious economic issue due to the mortality of lactating females, cost of treatment, reduced milk yield and quality [1,2], as well as a public health concern associated with risk of consumer food poisoning [3,4]. Several pathogens can cause mastitis in small ruminants; however, species of staphylococci are the most frequently isolated microorganisms from goat and sheep milk [2,5–8].

Staphylococcus aureus is one of the main pathogens associated with mastitis in small ruminants [9]. Incidence of clinical mastitis in sheep due to this bacterium may reach 20% with a mortality rate between 25% and 50%, and the affected mammary halves in surviving animals are frequently destroyed. Chronic mastitis may cause a 25 to 30% reduction in milk yield from the affected udder [10].

Coagulase negative staphylococci (CNS), although not as virulent as *S. aureus*, often cause subclinical mastitis in small ruminants [5,11–13]. This type of infection, most times not detected by the farmer, clearly reduces milk production, also changing milk composition, indirectly impairing the milk product's properties [14]. CNS are the most prevalent pathogens of the mammary gland in goats and sheep with subclinical mastitis, affecting

60% to 80.7% in goats and 45% to 48% in sheep [1]. Other authors have reported as much as 70.1% of subclinical mastitis in sheep is caused by CNS [5].

Virulence factors are bacterial molecules that enhance their capacity to establish and to survive within the host and, thus, contribute to bring damage to the host. Staphylococci possess a wide array of virulence factors [15].

Coagulase enzyme acts on plasma fibrinogen to form fibrin clots that protect the microorganisms from phagocytosis and shelter them from other cellular and soluble host defence mechanisms. This enzyme, encoded by the *coa* gene, is commonly used to distinguish coagulase positive staphylococci (CPS), namely *S. aureus*, *S. intermedius*, and *S. pseudintermedius*, from CNS species [16]. Nevertheless, this gene has been found also in species known as CNS such as *S. epidermidis*, *S. chromogenes*, and *S. hominis* [17]. The *coa* gene has also recently been associated with biofilm production [18].

The staphylococcal nuclease is a thermostable nuclease encoded by the *nuc* gene [19], which hydrolyzes DNA and RNA in host cells, causing tissue destruction and spreading of staphylococci [20], also promoting the escape of microorganisms when retained by neutrophil extracellular traps (NETs), allowing the bacteria to evade this host defence mechanism [21,22]. For decades, the *nuc* gene has been considered the golden standard for *Staphylococcus aureus* identification and is still used presently [23–25]. However, the *nuc* gene has been detected in staphylococci of animal origin other than *S. aureus* [26]. Moreover, the *nuc* encoded staphylococcal thermonuclease is a biofilm inhibitor that degrades the environmental DNA (eDNA) associated with biofilm [27,28].

The production of biofilm is considered a major virulence factor that, besides protecting from host defence mechanisms, also shields bacteria against antimicrobial agents [29]. Furthermore, the persistence of biofilm-producing isolates in the dairy environment enhances the dispersal of virulence factors though the transfer of genetic material to other bacteria [30]. Biofilm major components are an exopolysaccharide matrix, proteins, and eDNA, along with the bacterial cells [31]. The exopolysaccharide, polysaccharide intercellular adhesin (PIA), is also a non-protein adhesin [32] assisting in bacterial adhesion to different surfaces, comprising the first critical event in the establishment of an infection [33]. Staphylococcal PIA is encoded by the *ica* operon [34], and biofilm-associated protein (Bap) is a surface protein connected to the cell wall encoded by the *bap* gene [35].

Antimicrobial resistance (AMR) is a major problem hampering the treatment of an ever increasing range of infections caused by bacteria [36]. Staphylococci resistance has been reported for different antimicrobials used for mastitis control in small ruminants [7,36–38]. Genes often described in *Staphylococcus* spp. isolated from the milk of small ruminants are *blaZ* and *mecA*, responsible for β-lactam resistance and *tetK* and *tetM*, accounting for tetracycline resistance [39–41]. The presence of resistant bacteria in contaminated food products may lead to the transfer of resistance genes to the indigenous microbiota in the human gut [42].

The aim of this study was to identify *Staphylococcus* species isolated from small ruminants' milk samples and investigate the presence of genes encoding virulence factors associated with both biofilm (*coa*, *nuc*, *bap*, *icaA*, and *icaD*) and antimicrobial resistance to β-lactams (*blaZ*, *mecA*, and *mecC*) and tetracyclines (*tetK* and *tetM*).

2. Results and Discussion

2.1. Bacteriological Results

From the 646 milk samples collected from goats (508) and sheep (138), bacteriological cultures resulted positive in 191 samples: 131 goat milk and 60 sheep milk. A total of 137 staphylococcal isolates were recovered, of which 86 were isolated from goat and 51 from sheep milk samples.

2.2. Staphylococci Identification

Excellent (96 to 99% probability) and very good (93 to 95% probability) identification was observed for most *Staphylococcus*. Unidentified isolates and isolates with low discrimination results were confirmed by 16S rRNA gene sequencing.

Concerning goat milk samples, four *S. aureus*, one *Staphylococcus* sp., and 12 different CNS species were found: *S. caprae* (25), *S. chromogenes* (10), *S. epidermidis* (11), *S. simulans* (8), *S. warneri* (7), *S. capitis* (4), *S. lentus* (4), *S. hominis* (4), *S. hyicus* (3), *S. auricularis* (2), *S. haemolyticus* (2), and *S. equorum* (1).

On the other hand, 31 *S. aureus* and seven different CNS species were recovered from sheep milk samples: *S. chromogenes* (9), *S. epidermidis* (3), *S. auricularis* (2), *S. haemolyticus* (2), *S. simulans* (2), *S. lentus* (1), and *S. rostri* (1). *Staphylococcus rostri* has only been seldom isolated from the milk of a sheep with subclinical mastitis [43,44].

In the CNS group, *S. caprae* was the most found species and was isolated only from goat's milk samples. It is a commensal organism that prevails in the skin of the goat udder [19] This species is most commonly found in cases of goat mastitis [37,45–47], but it was also isolated from sheep [5,48], buffalo [17], and cow's milk [49].

In this study, other *Staphylococcus* species were only isolated from goats: *S. warneri*, *S. capitis*, *S. hominis*, *S hyicus*, and *S. equorum*. This was probably because the sheep sampling was smaller, since all these species have been isolated before from sheep milk by several other authors [44].

2.3. Biofilm Production

Of the 137 *Staphylococcus* isolates analyzed, 103 were biofilm producers (75%). Biofilm-forming isolates belong to the following species: *S. aureus* (29/35), *S. caprae* (22/25), *S. chromogenes* (12/19), *S. epidermidis* (11/14), *S. warneri* (7/7), *S. simulans* (6/10), *S. auricularis* (4/4), *S. capitis* (3/4), *S. lentus* (3/5), *S. haemolyticus* (2/4), *S. hominis* (2/4), *S. equorum* (1/1), and *Staphylococcus* sp. (1/1). All *S. epidermidis* goat isolates were found to produce biofilm in the present study, in accordance with the findings of others authors that reported *S. epidermidis* as the most commonly found species in biofilm-associated human infections [50]. However, none of the sheep *S. epidermidis* isolates were biofilm producers. In fact, other studies had already reported only 8% of biofilm-producing isolates among sheep mastitis *S. epidermidis* [51].

2.4. Genes Associated to Biofilm

We investigated the presence of *coa* and *nuc* genes in all 137 staphylococcal isolates, mainly for identification purposes and due to historical reasons. In fact, the ability of a strain to produce coagulase, encoded by the *coa* gene, is the basis of the primary classification of staphylococci in coagulase-positive or coagulase-negative [16].

All *S. aureus* isolates (35) harbored the *coa* gene, as well as isolate B200E1, not identified to the species level. Based on this result, this isolate was probable also *S. aureus*. Therefore, the 101 *Staphylococcus* isolates not carrying the *coa* gene were confirmed as CNS. Furthermore, in the present study, different amplicons of the *coa* gene with band sizes ranging from 400 to 900 bp were detected (Figure 1), as already reported by others [52–55]. In fact, the *coa* gene also has a discriminatory power between isolates because of the heterogeneity of its 3′ variable region containing 81-bp tandem short sequence repeats (SSR) [56–58].

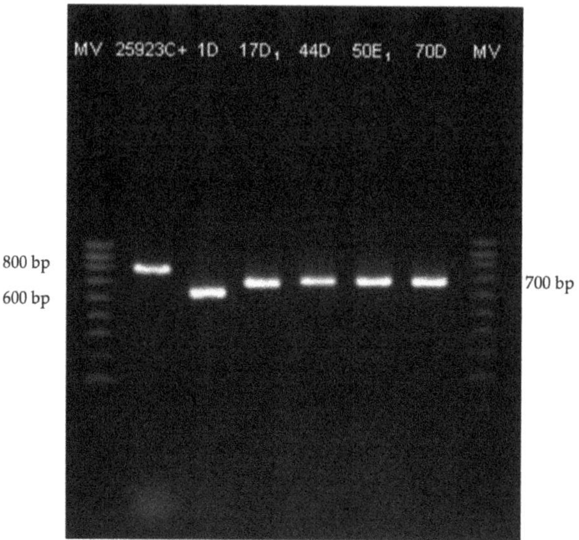

Figure 1. Agarose gel electrophoresis of *S. aureus coa* gene PCR products. NZYDNA Ladder V (200–1000 bp) (NZYTech, Lisbon, Portugal).

The *nuc* gene was detected in 67 out of 137 isolates (48.9%), of which only 35 were *S. aureus*. The other *nuc* positive isolates included: *S. chromogenes* (8), *S. warneri* (4), *S. auricularis* (3), *S. caprae* (3), *S. hyicus* (3), *S. lentus* (3), *S. epidermidis* (2), *S. simulans* (2), *S. capitis* (1), *S. haemolyticus* (1), *S. hominis* (1), and *Staphylococcus* sp. (1). Furthermore, an association was found between the *Staphylococcus* species and the presence of the *nuc* gene (χ^2 = 70.968, df = 14, $p < 0.001$). In fact, all *S. aureus* harbor the *nuc* gene, while most CNS (70/101) do not. However, the *nuc* gene was also detected in more than 50% of the isolates in some CNS species: *S. warneri* (4/7), *S. lentus* (3/5), *S. auricularis* (3/4), and *S. hyicus* (3/3).

The presence of the *nuc* gene was used in the past to identify *S. aureus* [23,25]. The *nuc* gene is present in most *S. aureus* isolates; however, some isolates not carrying this gene have been described [59,60]. Moreover, the *nuc* gene has also been detected in other species of *Staphylococcus*, both CPS and CNS [61,62].

For the detection of the biofilm production genes, *bap*, *icaA*, and *icaD*, the 44 *nuc*-positive biofilm-producing isolates were selected. *nuc*-positive biofilm-producing staphylococci and biofilm-associated genes are shown in Table 1.

Table 1. *nuc*-positive biofilm-producing staphylococcal isolates and biofilm-associated genes.

Isolate	Origin	Animal	Bacterial Species	*coa*	*nuc*	*bap*	*icaA*	*icaD*
1D	PT	goat	*S. aureus*	+	+	−	+	+
13D1	PT	goat	*S. warneri*	−	+	−	−	−
17D1	PT	goat	*S. aureus*	+	+	−	−	+
44D	PT	goat	*S. aureus*	+	+	−	+	+
47D2	PT	goat	*S. chromogenes*	−	+	+	−	−
50E1	PT	goat	*S. aureus*	+	+	−	+	+
54E1	PT	goat	*S. warneri*	−	+	+	−	−
54E2	PT	goat	*S. warneri*	−	+	−	+	−
55D1	PT	goat	*S. capitis*	−	+	−	−	−
60D2	PT	goat	*S. chromogenes*	−	+	+	−	−

Table 1. *Cont.*

Isolate	Origin	Animal	Bacterial Species	*coa*	*nuc*	*bap*	*icaA*	*icaD*
65D	PT	goat	*S. caprae*	−	+	−	−	−
70D	PT	sheep	*S. aureus*	+	+	−	−	+
71E	PT	sheep	*S. aureus*	+	+	−	−	−
72D	PT	sheep	*S. aureus*	+	+	−	−	+
72E	PT	sheep	*S. aureus*	+	+	−	−	+
83D	PT	sheep	*S. aureus*	+	+	−	−	−
B51E	BR	goat	*S. chromogenes*	−	+	−	−	−
B64	BR	goat	*S. chromogenes*	−	+	−	−	−
B76E	BR	goat	*S. chromogenes*	−	+	+	−	−
B101	BR	goat	*S. warneri*	−	+	−	+	−
B159D	BR	goat	*S. chromogenes*	−	+	+	−	−
B159E	BR	goat	*S. chromogenes*	−	+	+	−	−
B190D	BR	goat	*S. auricularis*	−	+	−	−	−
B209D2	BR	goat	*S. simulans*	−	+	+	−	−
B209E	BR	goat	*S. simulans*	−	+	−	−	−
B219D3	BR	sheep	*S. auricularis*	−	+	−	−	−
B219D5	BR	sheep	*S. aureus*	+	+	−	−	−
B223D	BR	sheep	*S. aureus*	+	+	−	−	−
B250D	BR	sheep	*S. auricularis*	−	+	+	−	−
CQ152E1	PT	sheep	*S. aureus*	+	+	−	+	+
CQ185D1	PT	sheep	*S. aureus*	+	+	−	+	+
CQ196E	PT	sheep	*S. aureus*	+	+	−	−	+
CQ201E	PT	sheep	*S. aureus*	+	+	−	−	+
CQ268D1	PT	sheep	*S. aureus*	+	+	−	−	+
CQ270E1	PT	sheep	*S. aureus*	+	+	−	−	−
CQ285D	PT	sheep	*S. aureus*	+	+	−	−	+
CQ286D	PT	sheep	*S. aureus*	+	+	−	−	+
CQ290D1	PT	sheep	*S. aureus*	+	+	−	−	+
CQ290D2	PT	sheep	*S. aureus*	+	+	−	−	+
CQ296D	PT	sheep	*S. aureus*	+	+	−	−	+
CQ335E	PT	sheep	*S. aureus*	+	+	−	−	−
CQ336E2	PT	sheep	*S. aureus*	+	+	−	−	+
CQ349D	PT	sheep	*S. aureus*	+	+	−	−	−
CQ354D	PT	sheep	*S. aureus*	+	+	−	−	+

PT-Portugal; BR-Brazil.

The *bap* gene was amplified in eight isolates: *S. chromogenes* (5), *S. auricularis* (1), *S. simulans* (1), and *S. warneri* (1). None of the *S. aureus nuc*-positive biofilm-producing isolates carries the *bap* gene. In fact, the *bap* gene has been reported mainly in *S. aureus* strains isolated from cattle [24,63,64]. However, Martins et al. [65] have detected the *bap* gene in four sheep milk *S. aureus* isolates. In our study, 8 out of 18 CNS *nuc*-positive biofilm-producing isolates harbored the *bap* gene. The *bap* gene encodes a cell wall associated protein named Bap (for biofilm associated protein), which enhances biofilm formation as it mediates bacterial primary attachment to abiotic surfaces and intercellular adherence [35]. Other studies have reported the presence of the *bap* gene in several CNS isolates [66].

The presence of the *icaA* gene was detected in seven isolates: *S. aureus* (5) and *S. warneri* (2). On the other hand, the *icaD* gene was present in 19 *S. aureus* isolates. Furthermore, five *S. aureus* isolates carried both *icaA* and *icaD* genes simultaneously. Xu, Tan, Zhang, Xia, and Sun [59] detected the *icaD* gene in 20 out of 28 *S. aureus* bovine mastitis isolates, while it was not detected in any of the 76 CNS analyzed. The same authors reported the absence of the *icaA* gene in all analyzed staphylococcal isolates [59].

No isolate carrying the *bap* gene harbored the *ica* operon genes, as reported before by other authors [67]. However, Marques et al. [68] found one single bovine mastitis *S. aureus* isolate (out of 20) that simultaneously carried *bap*, *icaA*, and *icaD*.

None of the three biofilm-associated genes were detected in 14 of the *nuc*-positive biofilm-producing isolates: *S. aureus* (6) and CNS (8). Other authors have also reported the absence of *bap*, *icaA*, and *icaD* genes in biofilm-producing *S. aureus* [24,69,70]. Despite no association being found between the presence of the *nuc* gene and biofilm production, most biofilm-producing isolates harbored the *nuc* gene (53.4%), while it was only detected in about 35% of the non-producers. Nevertheless, Kiedrowski, Kavanaugh, Malone, Mootz, Voyich, Smeltzer, Bayles, and Horswill [28] described an inverse correlation between Nuc thermonuclease activity and biofilm formation and confirmed the important role for eDNA in the *S. aureus* biofilm matrix.

Apparently, CNS produce biofilm mainly via Bap, as already suggested by Zuniga et al. [71], who found the *bap* gene to be more frequently present in CNS when compared to CPS.

Meanwhile, most *S. aureus* seem to form biofilm through PIA since they harbor the *icaA* and *icaD* genes. Other authors have reported that a low prevalence of the *bap* gene in *S. aureus* indicates that the *ica* operon-dependent mechanism may be the main responsible for the adhesion and biofilm formation in this species [68]. Notwithstanding, it has been reported that biofilm synthesis in *S. aureus* can also be encoded by the *bap* gene [72].

Other biofilm formation mechanisms in staphylococci not harboring the classical biofilm-production genes, *bap*, *icaA*, and *icaD*, need to be explored. Furthermore, some of the isolates not carrying *bap*, *icaA*, and *icaD* also did not harbor the *coa* gene, which has been reported as associated with biofilm formation [18]. However, the *nuc* gene might be an important factor to consider since all 44 isolates were biofilm producers and harbored the *nuc* gene, although Nuc has been referred to as a biofilm inhibitor [27,28].

2.5. Antimicrobial Resistance

Out of 137 staphylococcal isolates analyzed for antimicrobial susceptibility, 15 were multidrug resistant, 36 were non-susceptible to two antimicrobial categories, and 61 to one antimicrobial category, according to the classification proposed by Magiorakos et al. [73]. Moreover, no antimicrobial resistances were detected in 24 staphylococcal isolates.

Staphylococci isolated from milk from small ruminants with mastitis are known for their multiresistance [74]. In this work, the multidrug resistant (MDR) isolates belonged to the following species: *S. aureus* (8), *S. lentus* (3), *S. chromogenes* (2), *S. caprae* (1), and *S. warneri* (1). Contrarily, Taponen and Pyorala [75] reported that multiresistance was more common in CNS than in *S. aureus* from bovine mastitis.

Susceptibility patterns of CPS and CNS isolates are shown in Figure 2. For most antimicrobials tested, a higher percentage of resistant isolates was observed among CNS when compared to CPS. Vasileiou et al. [76] also reported more resistant CNS isolates than *S. aureus*. However, mastitis caused by CNS responds much better to antimicrobial treatment than *S. aureus* mastitis [75].

Staphylococcal isolates were mainly non-susceptible to streptomycin (50/137), penicillin (38/137), ampicillin (34/137), lincomycin (33/137), oxacillin (22/137), cloxacillin (21/137), and tetracycline (17/137), as previously reported [77] (Figure 2). Moreover, most CPS isolates were non-susceptible to streptomycin and lincomycin. On the other hand, CNS isolates were mostly non-susceptible to the β-lactams and tetracyclines.

In addition, an association was found between *Staphylococcus* species and antimicrobial resistance to penicillin (χ^2 = 45.981, df = 14, $p < 0.001$), ampicillin (χ^2 = 48.327, df = 14, $p < 0.001$), streptomycin (χ^2 = 137.705, df = 28, $p < 0.001$), lincomycin (χ^2 = 156.536, df = 28, $p < 0.001$), cephalexin (χ^2 = 57.219, df = 28, $p < 0.05$), and tetracycline (χ^2 = 51.626, df = 28, $p < 0.05$). Regarding the results shown by the correspondence analysis, most *S. caprae* and *S. capitis* isolates were resistant to penicillin and ampicillin, while all other staphylococci were mostly susceptible to these antimicrobials (Figure 3). Most *S. aureus* isolates exhibited an intermediate susceptibility pattern to streptomycin and lincomycin [78]. Additionally, all *S. hyicus* isolates were resistant to streptomycin, while *S. lentus* and *S. rostri* were resistant to lincomycin (Figure 3).

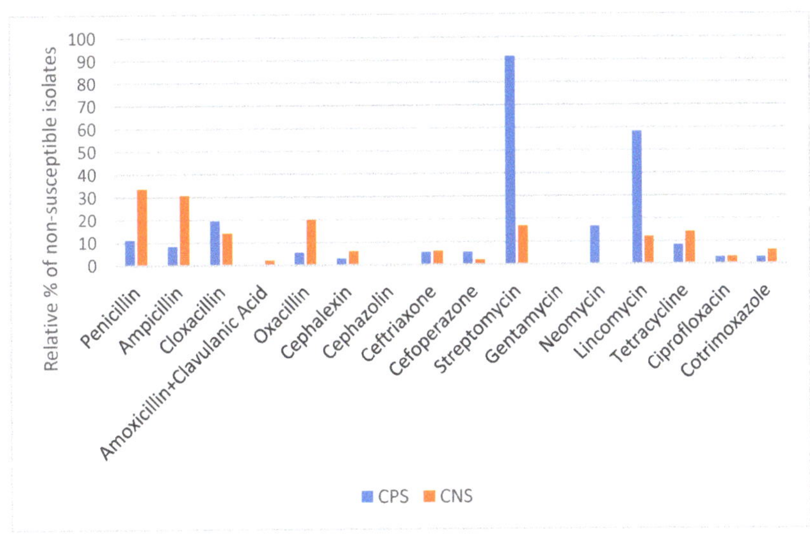

Figure 2. Susceptibility patterns of CPS (*n* = 36) and CNS (*n* = 101) isolates to antimicrobials.

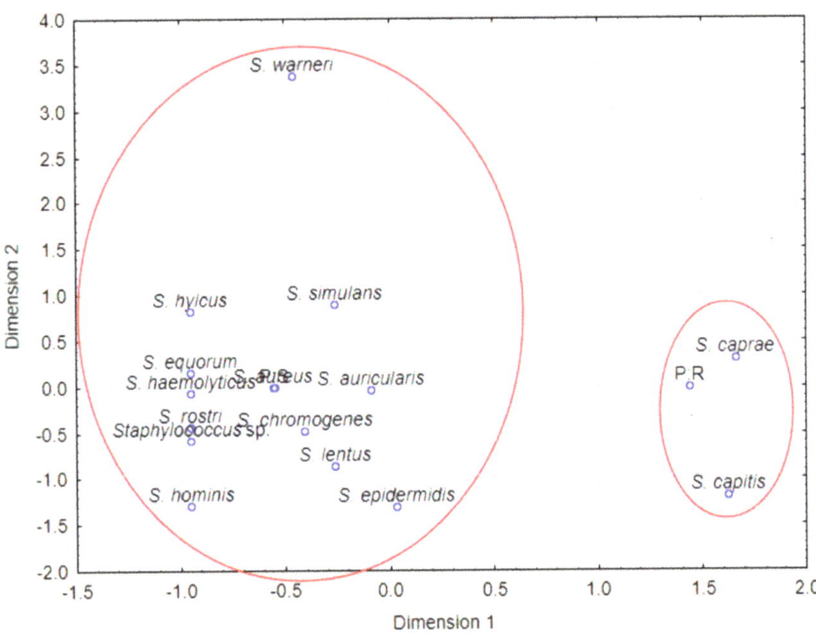

Figure 3. CA biplots of the relationship between bacterial species and tolerance to the antimicrobials penicillin (*P*), ampicillin (AMP), streptomycin (S), and lincomycin (MY).

No staphylococci resistant to cefazolin and gentamycin were identified. Moreover, no non-susceptible *S. aureus* isolates were found to amoxicillin + clavulanic acid. A number of CNS isolates, although resistant to penicillinase-labile penicillins, were susceptible to amoxicillin + clavulanic acid, which was expected due to the inhibitory action of clavulanic acid against β-lactamases [79]. Regarding CNS isolates, none were found to be resistant to neomycin.

One *S. aureus* and one CPS *Staphylococcus* sp. were found to be resistant to oxacillin, while CNS oxacillin resistant isolates belonged to eight species: *S. chromogenes* (5), *S. caprae* (4), *S. lentus* (3), *S. simulans* (3), *S. epidermidis* (2), *S. auricularis* (1), *S. hominis* (1), and *S. warneri* (1). Other authors previously reported the presence of methicillin resistant coagulase-negative staphylococci (MR-CNS) [80,81].

Regarding tetracycline, most *S. aureus* isolates (32/35) were susceptible, while non-susceptible isolates belonged to the following CNS species: *S. caprae* (4), *S. haemolyticus* (3), *S. lentus* (2), *S. capitis* (1), *S. hominis* (2), *S. rostri* (1), and *S. warneri* (1). Tetracycline has been widely used in veterinary medicine, and other studies have reported a higher percentage of resistant isolates: 42.8% [82] and 28.9% [45]. On the contrary, our results show a relatively low percentage of non-susceptible isolates (12.4%). In recent years, there has been an abusive use of more recent antimicrobial molecules, such as cephalosporins and quinolones, that may justify the observed reversal in the patterns of resistance to tetracyclines. To avoid the use of critically important antimicrobials for human medicine, tetracyclines, gentamycin, or cefazolin, a first-generation cephalosporin, may be an option for the control of mastitis in small ruminants. However, there should be a tight control over the development of antimicrobial resistances.

Interestingly, an association between resistance to some antibiotics and animal species was found: penicillin ($\chi^2 = 26.931$, df = 1, $p < 0.001$), ampicillin ($\chi^2 = 26.818$, df = 1, $p < 0.001$), oxacillin ($\chi^2 = 6.241$, df = 1, $p < 0.05$), streptomycin ($\chi^2 = 26.231$, df = 2, $p < 0.001$), and lincomycin ($\chi^2 = 20.831$, df = 2, $p < 0.001$). For example, isolates from goats (G) were more resistant than sheep (S) isolates to β-lactams, penicillin (G-43%; S-2%), ampicillin (G-39%; S- = 0%), and oxacillin (G-22%; S-6%). These differences might be due to different management systems, as suggested by Barrero-Domínguez et al. [45], who reported sheep and goat staphylococcal isolates with the same pulsotypes to exhibit distinct resistance patterns.

2.6. Antimicrobial Resistance Genes

The 44 biofilm producing isolates were selected for the detection of antimicrobial resistance genes involved in the resistance to β-lactams and tetracyclines, namely, *blaZ*, *mecA*, *mecC*, *tetK*, and *tetM*. Table 2 shows the antimicrobial genes detected in each isolate, along with its antimicrobial resistance profile.

The *blaZ* gene was detected in 15 staphylococcal isolates belonging to the following species: *S. chromogenes* (7), *S. aureus* (3), *S. warneri* (2), *S. auricularis* (1), *S. caprae* (1), and *S. simulans* (1). Unexpectedly, nine penicillin-susceptible isolates harbor the *blaZ* gene, namely *S. chromogenes* (5), *S. warneri* (2), *S. auricularis* (1), and *S. simulans* (1). El Feghaly et al. [83] also reported penicillin-susceptible isolates harboring the *blaZ* gene and concluded that conventional methods for susceptibility testing such as Kirby-Bauer penicillin disk diffusion may not be reliable. According to CLSI [78], there may be rare isolates of staphylococci containing β-lactamase genes, which may result negative in phenotypic β-lactamase detection. Additionally, all isolates resistant to penicillin must be considered resistant to all penicillinase-labile penicillins [78].

No staphylococcal isolates harboring the *mecA* or *mecC* genes were detected, although two isolates were found to be non-susceptible to oxacillin and cloxacillin simultaneously, one only to oxacillin and seven to cloxacillin alone. According to the CLSI (2016), oxacillin disk diffusion testing is not reliable for detecting methicillin resistance, at least in *S. aureus*, and cefoxitin should be used for disk diffusion testing. However, Barrero-Domínguez, Luque, Galán-Relaño, Vega-Pla, Huerta, Román, and Astorga [45] also did not detect the *mecA* gene in a cefoxitin-resistant MRSA strain. Thus, other resistance mechanisms cannot be excluded, namely, overproduction of β-lactamase, modified penicillin-binding proteins, distinct SCCmec elements, as well as putative *mecA* mutations [84,85]. Furthermore, Becker et al. [86] have recently reported the presence of a *mecB* gene in a MRSA strain, negative for both *mecA* and *mecC* genes. However, concerning *mecC* detection in our study, we cannot conclude that the isolates with a negative PCR result did not harbor the *mecC* gene, since no positive control strain was available.

Table 2. *nuc*-positive biofilm-producing staphylococcal isolates, phenotypic resistance to selected antimicrobials and their associated antimicrobial resistance genes.

Isolate	Origin	Animal	Bacterial Species	P	AMP	OB	AMC	OXA	TET	*blaZ*	*mecA*	*mecC*	*tetK*	*tetM*
1D	PT	goat	*S. aureus*	R	R	R	S	S	S	+	-		-	-
13D1	PT	goat	*S. warneri*	S	S	S	S	S	S	-	-		-	-
17D1	PT	goat	*S. aureus*	R	R	R	S	S	R	+	-		-	+
44D	PT	goat	*S. aureus*	R	R	S	S	S	S	+	-		-	-
47D2	PT	goat	*S. chromogenes*	R	R	R	R	R	S	+	-	-	-	-
50E1	PT	goat	*S. aureus*	S	S	S	S	S	S	-	-		-	-
54E1	PT	goat	*S. warneri*	S	S	R	S	S	S	+	-		-	-
54E2	PT	goat	*S. warneri*	S	S	R	S	R	S	-	-	-	-	-
55D1	PT	goat	*S. capitis*	S	S	S	S	S	S	-	-		-	-
60D2	PT	goat	*S. chromogenes*	R	R	S	S	S	S	+	-		-	-
65D	PT	goat	*S. caprae*	R	R	S	S	S	S	+	-		-	-
70D	PT	sheep	*S. aureus*	S	S	S	S	S	S	-	-		-	-
71E	PT	sheep	*S. aureus*	S	S	R	S	S	S	-	-		-	-
72D	PT	sheep	*S. aureus*	S	S	S	S	S	S	-	-		-	-
72E	PT	sheep	*S. aureus*	S	S	S	S	S	S	-	-		-	-
83D	PT	sheep	*S. aureus*	S	S	S	S	S	S	-	-		-	-
B51E	BR	goat	*S. chromogenes*	S	S	S	S	S	S	+	-		-	-
B64	BR	goat	*S. chromogenes*	S	S	S	S	S	S	+	-		-	-
B76E	BR	goat	*S. chromogenes*	S	S	S	S	S	S	+	-		-	-
B101	BR	goat	*S. warneri*	S	S	S	S	S	R	+	-		-	-
B159D	BR	goat	*S. chromogenes*	S	S	S	S	S	S	+	-		-	-
B159E	BR	goat	*S. chromogenes*	S	S	S	S	S	S	+	-		-	-
B190D	BR	goat	*S. auricularis*	R	S	S	S	S	S	-	-		-	-
B209D2	BR	goat	*S. simulans*	S	S	S	S	S	S	-	-		-	-
B209E	BR	goat	*S. simulans*	S	S	S	S	S	S	+	-		-	-
B219D3	BR	sheep	*S. auricularis*	S	S	S	S	S	S	-	-		-	-
B219D5	BR	sheep	*S. aureus*	S	S	S	S	S	S	-	-		-	-
B223D	BR	sheep	*S. aureus*	S	S	S	S	R	S	-	-	-	-	-
B250D	BR	sheep	*S. auricularis*	S	S	S	S	S	S	+	-		-	-
CQ152E1	PT	sheep	*S. aureus*	S	S	S	S	S	S	-	-		-	-
CQ185D1	PT	sheep	*S. aureus*	S	S	S	S	S	S	-	-		-	-
CQ196E	PT	sheep	*S. aureus*	S	S	S	S	S	S	-	-		-	-
CQ201E	PT	sheep	*S. aureus*	S	S	S	S	S	S	-	-		-	-
CQ268D1	PT	sheep	*S. aureus*	S	S	R	S	S	S	-	-		-	-
CQ270E1	PT	sheep	*S. aureus*	S	S	R	S	S	S	-	-		-	-
CQ285D	PT	sheep	*S. aureus*	S	S	S	S	S	S	-	-		-	-
CQ286D	PT	sheep	*S. aureus*	S	S	S	S	S	S	-	-		-	-
CQ290D1	PT	sheep	*S. aureus*	S	S	R	S	S	S	-	-		-	-
CQ290D2	PT	sheep	*S. aureus*	S	S	S	S	S	S	-	-		-	-
CQ296D	PT	sheep	*S. aureus*	S	S	S	S	S	R	-	-		+	-
CQ335E	PT	sheep	*S. aureus*	S	S	S	S	S	S	-	-		-	-
CQ336E2	PT	sheep	*S. aureus*	S	S	S	S	S	S	-	-		-	-
CQ349D	PT	sheep	*S. aureus*	S	S	S	S	S	S	-	-		-	-
CQ354D	PT	sheep	*S. aureus*	S	S	S	S	S	S	-	-		-	-

Penicillin (P), ampicillin (AMP), cloxacillin (OB), amoxicillin + clavulanic acid (AMC), oxacillin (OXA), tetracyclines-tetracycline (TET).

An association was found between the resistance to penicillin ($\chi^2 = 11.650$, df = 1, $p < 0.05$) and ampicillin ($\chi^2 = 15.828$, df = 1, $p < 0.001$) and the presence of the antimicrobial resistance gene *blaZ*. The association between resistance to penicillin and ampicillin and the presence of the antimicrobial resistance gene *blaZ* has been reported before by other authors [87,88]. However, no association was detected between the resistance to oxacillin and cloxacillin and the presence of the antimicrobial resistance gene *mecA* for this subgroup of 44 isolates.

Only one *S. aureus* isolate carrying the *tetK* and another one carrying the *tetM* gene were identified. Both showed resistance to tetracycline. A *S. warneri* tetracycline-resistant isolate did not harbor either *tetK* or *tetM* (Table 3). El-Razik, Arafa, Hedia, and Ibrahim [82] found a *S. intermedius* isolate showing intermediate resistance to tetracycline, not harboring *tetK*, *tetL*, *tetM*, and *tetO* genes.

3. Materials and Methods

3.1. Milk Samples Collection and Bacteriological Analyses

A total of 328 small ruminants (258 goats and 70 sheep), belonging to 23 both traditional and industrial dairy farms in Portugal and Brazil, were used to collect 646 half-udder milk samples (508 from goats and 138 from sheep).

Milk samples were aseptically collected in a sterile bottle after the teat was carefully disinfected with 70% ethanol and the first flush was rejected. The samples were kept refrigerated and transported to the laboratory. Ten microliters of each milk sample were plated onto MacConkey agar (Oxoid, Hampshire, UK, CM0007) and onto blood agar (BA) (Oxoid, Hampshire, UK; CM0271 with 5% sheep blood) and incubated at 37 °C for 24 h to 48 h.

Colonies from BA were transferred to brain heart infusion agar (BHI) (Oxoid, Hampshire, UK, CM1136) and again incubated at 37 °C for 24h for primary identification of the *Staphylococcus* genus through morphological and biochemical characteristics, namely, colony morphology, Gram staining, and catalase reaction, according to Markey et al. [89].

Identification of the species level of all isolates was performed by automated compact system VITEK 2 (bioMérieux, Marcy l'Etoile, France) using GP ID cards following the manufacturer's instructions. Biochemical identification was confirmed by 16S rRNA gene sequencing whenever necessary, using the primers described previously [90].

3.2. Phenotypic Characterisation of Staphylococcal Isolates

3.2.1. Biofilm Detection

Biofilm production was evaluated following the procedures described by Merino et al. [91] with some modifications. In brief, isolates were grown overnight in trypticase soy broth (TSB) at 37 °C. This overnight culture was diluted 1:40 in TSB supplemented with 0.25% glucose, and 200 mL of this cell suspension was used to inoculate microplates. After 24 h of incubation at 37 °C, the microplates were washed three times with 200 μL H_2O, dried in an inverted position, and stained with 100 μL of 0.25% crystal violet for 2 to 3 min at room temperature. Afterwards, the microplates were rinsed again three times with H_2O, dried, the dye dissolved in 200 μL ethanol-acetone (80:20), and the absorbance measured at 620 nm. Each assay was performed in triplicate and repeated three times. *Staphylococcus epidermidis* ATCC 12,228 and ATCC 35,984 were used as negative and positive controls, respectively. A blank control was also used.

3.2.2. Antimicrobial Sensitivity Test

The antimicrobial sensitivity test (AST) was performed as described before [77] following the performance standard M02-A11 [92]. Resistance to 16 antimicrobials, belonging to six antimicrobial categories, according to Magiorakos et al. [73], was evaluated: (1) β-lactams-penicillin (P), ampicillin (AMP), cloxacillin (OB), amoxicillin + clavulanic acid (AMC), oxacillin (OXA), cephalexin (CL), cefazolin (KZ), ceftriaxone (CRO), cefoperazone (CFP); (2) aminoglycosides-streptomycin (S), gentamycin (CN), neomycin (N); (3) lincosamides-lincomycin (MY); (4) tetracyclines-tetracycline (TET); (5) fluoroquinolones-ciprofloxacin (CIP); and (6) folate pathway inhibitors-cotrimoxazole (sulfamides + trimethoprim) (STX).

For the interpretation of AST results, the CLSI clinical breakpoints M100-S25 were used [78]. Isolates showing intermediate resistance, now called "susceptible increased exposure" [93], were considered non-susceptible. Moreover, isolates resistant to three or more antimicrobial categories were considered multidrug resistant [73].

3.3. Molecular Characterisation of Staphylococcal Isolates

The presence of *coa* and *nuc* genes was investigated in all staphylococcal isolates. *nuc*-positive biofilm-producing isolates were selected for the detection of the biofilm production genes, *bap*, *icaA*, and *icaD*, and the antimicrobial resistance genes *blaZ*, *mecA*, *tetK*, and *tetM*. The presence of the *mecC* gene was investigated only for *nuc*-positive biofilm-producing isolates, which were simultaneously resistant to oxacillin and cloxacillin and did not harbor the *mecA* gene.

3.3.1. Rapid DNA Extraction

Total DNA was extracted as described previously [94]. Bacterial cultures were grown for 24 h in BHI (Oxoid, Hampshire, UK, CM1136). After this period, they were transferred to microtubes with 200 μL of ultrapure water and centrifuged at $12,000 \times g$ for two minutes. Two hundred microliters of sterile saline solution (8.5%) were added to the pellet and centrifuged again at $12,000 \times g$ for two minutes. Subsequently, 100 μL of 0.05 M NaOH was added to the pellet and boiled for four minutes, then placed immediately on ice. Afterwards, 600 μL of ultrapure water was added to the microtubes and centrifuged at $4000 \times g$ for three minutes. Subsequently, 400 μL were transferred to a new microtube and stored at $-20\ °C$ until use.

3.3.2. PCR Amplification

All amplifications were done in a PTC1148C-MJ Mini thermocycler (BioRad, Hercules, CA, USA).

Amplified DNA fragments were stained with 1X Red Gel (Biotium, Fremont, CA, USA) and run on 1.5% (*w/v*) agarose gels with 0.5X TBE (Tris-borate-EDTA) buffer. Different NZYDNA Ladders (NZYtech, Lisbon, Portugal) were used as molecular weight markers, depending on the size of the PCR products.

Agarose gels were photographed under ultraviolet light using the Gel Doc XR+ Gel Documentation System (BioRad Universal Hood II, Philadelphia, PA, USA).

For all PCR amplifications, 50 μL PCR reactions were prepared with 5 μL of DNA template, 1 U GoTaq DNA polymerase (Promega, Madison, WI, USA), 1X Green Go Taq Flexi buffer (Promega, WI, USA), 1.5 mM $MgCl_2$ (Promega, WI, USA), 0.2 mM each dNTP (VWR, part of Avantor, Radnor, PA, USA), and 15 pmol each primer (STAB VIDA, Caparica, Portugal). Specific and individual modifications or optimizations were done whenever necessary.

The primers used for amplification of the different genes are listed in Table 3.

Table 3. Primer sequences for amplification of the different genes.

Gene	Primer	Sequence	Reference
coa	coa-F coa-R	5′ ATA GAG ATG CTG GTA CAG G 3′ 5′ GCT TCC GAT TGT TCG ATG C 3′	[55]
coa	coa2-F coa2-R	5′ TA CTC AAC CGA CGA CAC CG 3′ 5′ GAT TTT GGA TGA AGC GGA TT 3′	[54]
nuc	nuc-F nuc-R	5′ GCG ATT GAT GGT GAT ACG GTT 3′ 5′ AGC CAA GCC TTG ACG AAC TAA AGC 3′	[95]
bap	bap-F bap-R	5′ CCC TAT ATC GAA GGT GTA GAA TTG CAC 3′ 5′ GCT GTT GAA GTT AAT ACT GTA CCT GC 3′	[35]
icaA	icaA-F icaA-R	5′ CCT AAC TAA CGA AAG GTA G 3′ 5′ AAG ATA TAG CGA TAA GTG C 3′	[96]
icaD	icaD-F icaD-R	5′ AAA CGT AAG AGA GGT GG 3′ 5′ GGC AAT ATG ATC AAG ATA C 3′	[96]
blaZ	blaZ-F blaZ-R	5′ AAG AGA TTT GCC TAT GCT TC 3′ 5′ GCT TGA CCA CTT TTA TCA GC 3′	[97]

Table 3. *Cont.*

Gene	Primer	Sequence	Reference
mecA	mecA-F	5′ AAA ATC GAT GGT AAA GGT TGG C 3′	[98]
	mecA-R	5′ AGT TCT GCA GTA CCG GAT TTG C 3′	
mecC	mecC-F	5′ GAA AAA AAG GCT TAG AAC GCC TC 3′	[99]
	mecC-R	5′ GAA GAT CTT TTC CGT TTT CAG C 3′	
tetK	tetK-F	5′ GTA GCG ACA ATA GGT AAT AGT 3′	[59]
	tetK-R	5′ TAG TGA CAA TAA ACC TCC TA 3′	
tetM	tetM-F	5′ AGT GGA GCG ATT ACA GAA 3′	[59]
	tetM-R	5′ CAT ATG TCC TGG CGT GTC TA 3′	

For the detection of the *coa* gene, different primer sequences were used. *Staphylococcus aureus* ATCC 25,923 was used as positive control. The first pair of primers, coa-F and coa-R, amplified a 676 bp fragment [55]. The amplification program was as follows: 3 min at 95 °C, and 35 cycles of 30 s at 94 °C, 30 s at 55 °C, 30 s at 72 °C, and finally, 5 min at 72 °C. The second pair of primers, coa2-F and coa2-R, amplified a fragment of 1517 bp [54]. The amplification program comprised an initial denaturation of 45 s at 94 °C, followed by 29 cycles at 94 °C for 20 s, 55 °C for 1 min, and 72 °C for 90 s, and a final extension step of 2 min at 72 °C.

For the amplification of the *nuc* gene, primers nuc-F and nuc-R, amplifying a 267 bp DNA fragment, were used [95]. *S. aureus* ATCC 25,923 was used as positive control and *S. epidermidis* ATCC 12,228 as negative control. The amplification program was the following: 5 min at 94 °C, followed by 37 cycles, consisting of 94 °C for 1 min, 55 °C for 30 s, and 72 °C for 30 s, ending with a final extension step at 72 °C for 7 min.

For detecting the *bap* gene, primers bap-F and bap-R were used for the amplification of a 971 bp fragment [35]. No positive control strain was available. The amplification program was as follows: 94 °C for 2 min, followed by 35 cycles of 94 °C for 45 s, 56.5 °C for 45 s, and 72 °C for 50 s, and finally, 72 °C for 5 min.

Primers icaA-F and icaA-R were used for the amplification of a 1315 bp fragment of the *icaA* gene [96]. *S. epidermidis* ATCC 35,984 was used as positive control. The following amplification program was used: 92 °C for 5 min, followed by 30 cycles of 92 °C for 45 s, 49 °C for 45 s, and 72 °C for 1 min, and a final extension step of 7 min at 72 °C.

For the *icaD* gene, primers icaD-F and icaD-R were used to amplify a 381 bp fragment [96]. *S. epidermidis* ATCC 35,984 was used as positive control. The same amplification program as for *icaA* was used, except for the extension step within the cycles, which was 72 °C for 30 s.

The presence of the *blaZ* gene was detected using primers blaZ-F and blaZ-R, which amplified a 517 bp fragment [97]. *Staphylococcus aureus* ATCC 29,213 was used as positive control and *S. aureus* ATCC 25,923 as negative control [100]. The amplification program was as follows: 94 °C for 4 min, followed by 37 cycles of 94 °C for 1 min, 50.5 °C for 30 s, and 72 °C for 30 s, and finally, 72 °C for 5 min [97].

To detect the *mecA* gene, primers mecA-F and mecA-R were used to amplify a fragment of 532 bp [98]. *Staphylococcus epidermidis* ATCC 35,984 was used as positive control [101] and *S. aureus* ATCC 25,923 as negative control [102]. The following amplification program was used: 94 °C for 2 min, followed by 29 cycles of 94 °C for 30 s, 55 °C for 30 s, and 72 °C for 30 s, and a final extension of 5 min at 72 °C.

Primers mecC-F and mecC-R were used to amplify a 138 bp fragment [99]. No positive control strain was available. The following amplification program was used: 95 °C for 2 min, followed by 30 cycles of 94 °C for 30 s, 50 °C for 30 s, and 72 °C for 30 s, and a final extension of 10 min at 72 °C.

Primers tetK-F and tetK-R were used to amplify a 360 bp fragment of the *tetK* gene [59]. No positive control strain was available. For the amplification of the *tetM* gene, tetM-F and tetM-R were used to amplify a fragment of 158 bp [59]. No positive control strain was available. The amplification program for both *tet* genes was: 94 °C for 2 min, followed by 29 cycles of 94 °C for 30 s, 55 °C for 30 s, and 72 °C for 30 s, with a final step of 5 min at 72 °C.

3.4. Data Analysis

The chi-square test of association was used: to assess the relationship between the presence of the *nuc* gene with *Staphylococcus* species; to investigate if the presence of the *nuc* gene was associated with biofilm production; to check if the resistance to antimicrobials was associated with bacterial species and with the animal species from which these were isolated. For the abovementioned analyses, all 137 isolates were considered.

For the subgroup of 44 *nuc*-positive biofilm-producing isolates, the chi-square test of association was performed to evaluate the putative relationship between phenotypic resistance to antimicrobials and the presence of four resistance genes.

Multiple correspondence analysis (MCA) was used as an exploratory data analysis technique to detect a structure in the relationships between bacterial species and resistance to selected antimicrobials, divided either into two (susceptible and resistant) or three classes (susceptible, intermediate, and resistant), depending on the antimicrobial.

All statistical analyses were performed using the software STATISTICA Version 12 (StatSoft, Inc., Tulsa, OK, USA).

4. Conclusions

Mastitis aetiology showed to be diverse in the two small ruminant species studied. The most abundant species was *S. caprae*, which, however, was only present in goats.

The *nuc* gene was detected in 67 isolates, of which only 35 were *S. aureus*. Most CNS did not harbor this gene; however, it was detected in more than 50% of *S. warneri*, *S. lentus*, *S. auricularis*, and *S. hyicus*. Although many studies still consider the *nuc* gene as the sole character to identify *S. aureus*, our results have clearly demonstrated that this gene is insufficient, because it is present in numerous staphylococcal isolates other than *S. aureus*.

Most staphylococci were biofilm producers. The *bap* gene was only detected in CNS, while *ica* operon genes were mainly detected in *S. aureus* isolates, suggesting that CNS produce biofilm mainly via Bap, and most *S. aureus* form biofilm through PIA. Furthermore, biofilm-producing staphylococcal isolates not harboring the classical biofilm-production genes *bap*, *icaA*, and *icaD* carry the *nuc* gene. Therefore, the role of the Nuc thermonuclease in staphylococci biofilm formation needs to be further investigated.

Antimicrobial resistance seems to be a growing concern in the treatment of sheep and goat mastitis, with only a low number of isolates (18%) not showing any antimicrobial resistances. Furthermore, CNS were generally more resistant to antimicrobials than CPS. Additionally, an association between animal species and resistance to some antimicrobials was found, suggesting different managing systems for the two species.

All staphylococcal isolates were susceptible to cefazolin and gentamycin. Furthermore, all *S. aureus* isolates were shown to be susceptible to amoxicillin + clavulanic acid and most (32/35) to tetracycline. The use of these antimicrobials to control mastitis may be encouraged to avoid the use of others critically important for human medicine that are currently being used, such as third generation cephalosporins and quinolones. Nevertheless, antimicrobial susceptibility tests cannot be neglected, as the development of resistant strains is always a problem.

Regarding antimicrobial resistance genes, nine penicillin-susceptible isolates exhibited the *blaZ* gene, highlighting the poor reliability of conventional methods for susceptibility testing. Moreover, no staphylococcal isolates harboring the *mecA* or *mecC* genes were detected among those found to be non-susceptible to oxacillin. Hence, other methicillin resistance mechanisms need to be explored.

Author Contributions: Conceptualization, M.L. and M.C.Q.; methodology, M.L. and M.C.Q.; validation, M.L. and M.C.Q.; formal analysis, M.L. and M.C.Q.; investigation, N.C.A.; resources, M.L. and M.C.Q.; writing—original draft preparation, N.C.A.; writing—review and editing, M.L. and

M.C.Q.; visualization, N.C.A., M.L. and M.C.Q.; supervision, M.L., M.M.C. and M.C.Q.; project administration, M.L. and M.C.Q.; funding acquisition, M.L. and M.C.Q. All authors have read and agreed to the published version of the manuscript.

Funding: This work was funded by National Funds through FCT-Foundation for Science and Technology under Project UIDB/05183/2020. Nara Andrade was funded by CNPq through a PhD grant (249398/2013-3-GDE), from Brazil.

Institutional Review Board Statement: Not applicable.

Informed Consent Statement: Not applicable.

Acknowledgments: The authors acknowledge Guilhermina Pias for her technical support.

Conflicts of Interest: The authors declare no conflict of interest. The funders had no role in the design of the study; in the collection, analyses, or interpretation of data; in the writing of the manuscript, or in the decision to publish the results.

References

1. Olechnowicz, J.; Jaśkowski, J.M. Mastitis in small ruminants. *Med. Weter.* **2014**, *70*, 67–72.
2. Gelasakis, A.I.; Mavrogianni, V.S.; Petridis, I.G.; Vasileiou, N.G.C.; Fthenakis, G.C. Mastitis in sheep—The last 10 years and the future of research. *Vet. Microbiol.* **2015**, *181*, 136–146. [CrossRef]
3. De Buyser, M.-L.; Dufour, B.; Maire, M.; Lafarge, V. Implication of milk and milk products in food-borne diseases in France and in different industrialised countries. *Int. J. Food Microbiol.* **2001**, *67*, 1–17. [CrossRef]
4. Argudín, M.Á.; Mendoza, M.C.; Rodicio, M.R. Food poisoning and *Staphylococcus aureus* enterotoxins. *Toxins* **2010**, *2*, 1751–1773. [CrossRef] [PubMed]
5. Queiroga, M.C. Prevalence and aetiology of sheep mastitis in Alentejo region of Portugal. *Small Rumin. Res.* **2017**, *153*, 123–130. [CrossRef]
6. Abdalhamed, A.M.; Zeedan, G.S.G.; Abou Zeina, H.A.A. Isolation and identification of bacteria causing mastitis in small ruminants and their susceptibility to antibiotics, honey, essential oils, and plant extracts. *Vet. World* **2018**, *11*, 355–362. [CrossRef] [PubMed]
7. Ayis, A.A.E.; Fadlalla, E. Antibiotic Susceptibility of Major Bacteria Cause Ovine Mastitis in River Nile State, Sudan. *Imp. J. Interdiscip. Res.* **2017**, *3*, 908–916.
8. Salaberry, S.R.S.; Saidenberg, A.B.S.; Zuniga, E.; Gonsales, F.F.; Melville, P.A.; Benites, N.R. Microbiological analysis and sensitivity profile of *Staphylococcus* spp. in subclinical mastitis of dairy goats. *Arq. Bras. Med. Veterinária Zootec.* **2016**, *68*, 336–344. [CrossRef]
9. Bergonier, D.; Sobral, D.; Feßler, A.T.; Jacquet, E.; Gilbert, F.B.; Schwarz, S.; Treilles, M.; Bouloc, P.; Pourcel, C.; Vergnaud, G. *Staphylococcus aureus* from 152 cases of bovine, ovine and caprine mastitis investigated by Multiple-locus variable number of tandem repeat analysis (MLVA). *Vet. Res.* **2014**, *45*, 97. [CrossRef] [PubMed]
10. Constable, P.D.; Hinchcliff, K.W.; Done, S.H.; Grünberg, W. *Veterinary Medicine: A Textbook of the Diseases of Cattle, Horses, Sheep, Pigs, and Goats*, 11th ed.; Elsevier: St. Louis, MO, USA, 2017; pp. 1904–2001. ISBN 9780-7020-5246-8.
11. Dore, S.; Liciardi, M.; Amatiste, S.; Bergagna, S.; Bolzoni, G.; Caligiuri, V.; Cerrone, A.; Farina, G.; Montagna, C.O.; Saletti, M.A.; et al. Survey on small ruminant bacterial mastitis in Italy, 2013–2014. *Small Rumin. Res.* **2016**, *141*, 91–93. [CrossRef]
12. Mishra, A.K.; Sharma, N.; Singh, D.D.; Gururaj, K.; Abhishek; Kumar, V.; Sharma, D.K. Prevalence and bacterial etiology of subclinical mastitis in goats reared in organized farms. *Vet. World* **2010**, *11*, 20–24. [CrossRef] [PubMed]
13. Ruiz Romero, R.A. Identification of and antimicrobial resistance in bacteria causing caprine mastitis in three states and a city in Central Mexico under manual and mechanical milking conditions. *J. Dairyvet. Anim. Res.* **2018**, *7*. [CrossRef]
14. Leitner, G.; Rovai, M.; Merin, U. Clinical and subclinical intrammamay infection caused by coagulase negative staphylococci negatively affect milk yield and its quality in dairy sheep. *Small Rumin. Res.* **2019**, *180*, 74–78. [CrossRef]
15. Madigan, M.T.; Stahl, D.A.; Sattley, W.M.; Buckley, D.H.; Bender, K.S. *Brock Biology Of Microorganisms, Global Edition*; Pearson Education Limited: London, UK, 2018.
16. Murray, P.R.; Rosenthal, K.S.; Pfaller, M.A. *Medical Microbiology*, 9th ed.; Elsevier: Amsterdam, The Netherlands, 2020.
17. Pizauro, L.J.L.; de Almeida, C.C.; Soltes, G.A.; Slavic, D.; de Ávila, F.A.; Zafalon, L.F.; MacInnes, J.I. Short communication: Detection of antibiotic resistance, *mecA*, and virulence genes in coagulase-negative *Staphylococcus* spp. from buffalo milk and the milking environment. *J. Dairy Sci.* **2019**, *102*, 11459–11464. [CrossRef]
18. Zapotoczna, M.; McCarthy, H.; Rudkin, J.K.; O'Gara, J.P.; O'Neill, E. An Essential Role for Coagulase in *Staphylococcus aureus* Biofilm Development Reveals New Therapeutic Possibilities for Device-Related Infections. *J. Infect. Dis.* **2015**, *212*, 1883–1893. [CrossRef] [PubMed]
19. Schleifer, K.-H.; Bell, J.A. Family VIII-Staphylococcaceae *fam. nov.* In *Bergey's Manual of Systematic Bacteriology*, 2nd ed.; Vos, P.D., Garrity, G.M., Jones, D., Krieg, N.R., Ludwig, W., Rainey, F.A., Schleifer, K.-H., Whitman, W.B., Eds.; Springer: New York, NY, USA, 2009; Volume 3, pp. 392–421.

20. Hu, Y.; Meng, J.; Shi, C.; Hervin, K.; Fratamico, P.M.; Shi, X. Characterization and comparative analysis of a second thermonuclease from *Staphylococcus aureus*. *Microbiol. Res.* **2013**, *168*, 174–182. [CrossRef] [PubMed]

21. Berends, E.T.M.; Horswill, A.R.; Haste, N.M.; Monestier, M.; Nizet, V.; Von Köckritz-Blickwede, M. Nuclease expression by *Staphylococcus aureus* facilitates escape from neutrophil extracellular traps. *J. Innate Immun.* **2010**, *2*, 576–586. [CrossRef] [PubMed]

22. Kenny, E.F.; Herzig, A.; Krüger, R.; Muth, A.; Mondal, S.; Thompson, P.R.; Brinkmann, V.; Bernuth, H.V.; Zychlinsky, A. Diverse stimuli engage different neutrophil extracellular trap pathways. *eLife* **2017**, *6*. [CrossRef]

23. Kateete, D.P.; Kimani, C.N.; Katabazi, F.A.; Okeng, A.; Okee, M.S.; Nanteza, A.; Joloba, M.L.; Najjuka, F.C. Identification of *Staphylococcus aureus*: DNase and Mannitol salt agar improve the efficiency of the tube coagulase test. *Ann. Clin. Microbiol. Antimicrob.* **2010**, *9*, 23. [CrossRef]

24. Torres, G.; Vargas, K.; Sánchez-Jiménez, M.; Reyes-Velez, J.; Olivera-Angel, M. Genotypic and phenotypic characterization of biofilm production by *Staphylococcus aureus* strains isolated from bovine intramammary infections in Colombian dairy farms. *Heliyon* **2019**, *5*, e02535. [CrossRef]

25. McClure, J.A.; Zaal DeLongchamp, J.; Conly, J.M.; Zhang, K. Novel Multiplex PCR Assay for Detection of Chlorhexidine-Quaternary Ammonium, Mupirocin, and Methicillin Resistance Genes, with Simultaneous Discrimination of *Staphylococcus aureus* from Coagulase-Negative Staphylococci. *J. Clin. Microbiol.* **2017**, *55*, 1857–1864. [CrossRef] [PubMed]

26. Gudding, R. Differentiation of staphylococci on the basis of nuclease properties. *J. Clin. Microbiol.* **1983**, *18*, 1098–1101. [CrossRef]

27. Mann, E.E.; Rice, K.C.; Boles, B.R.; Endres, J.L.; Ranjit, D.; Chandramohan, L.; Tsang, L.H.; Smeltzer, M.S.; Horswill, A.R.; Bayles, K.W. Modulation of eDNA release and degradation affects *Staphylococcus aureus* biofilm maturation. *PLoS ONE* **2009**, *4*, e5822. [CrossRef] [PubMed]

28. Kiedrowski, M.R.; Kavanaugh, J.S.; Malone, C.L.; Mootz, J.M.; Voyich, J.M.; Smeltzer, M.S.; Bayles, K.W.; Horswill, A.R. Nuclease modulates biofilm formation in community-associated methicillin-resistant *Staphylococcus aureus*. *PLoS ONE* **2011**, *6*, e26714. [CrossRef] [PubMed]

29. Flemming, H.-C.; Wingender, J.; Szewzyk, U.; Steinberg, P.; Rice, S.A.; Kjelleberg, S. Biofilms: An emergent form of bacterial life. *Nat. Rev. Microbiol.* **2016**, *14*, 563–575. [CrossRef] [PubMed]

30. Turchi, B.; Bertelloni, F.; Marzoli, F.; Cerri, D.; Tola, S.; Azara, E.; Longheu, C.M.; Tassi, R.; Schiavo, M.; Cilia, G.; et al. Coagulase negative staphylococci from ovine milk: Genotypic and phenotypic characterization of susceptibility to antibiotics, disinfectants and biofilm production. *Small Rumin. Res.* **2020**, *183*, 106030. [CrossRef]

31. Buttner, H.; Dietrich, M.; Rohde, H. Structural basis of *Staphylococcus epidermidis* biofilm formation: Mechanisms and molecular interactions. *Front. Cell. Infect. Microbiol.* **2015**, *5*, 14. [CrossRef]

32. Arciola, C.R.; Campoccia, D.; Ravaioli, S.; Montanaro, L. Polysaccharide intercellular adhesin in biofilm: Structural and regulatory aspects. *Front. Cell. Infect. Microbiol.* **2015**, *5*, 7. [CrossRef]

33. Heilmann, C. Adhesion Mechanisms of Staphylococci. *Adv. Exp. Med. Biol.* **2011**, *715*, 105–123. [CrossRef]

34. Otto, M. Staphylococcal Biofilms. *Microbiol. Spectr.* **2018**, *6*. [CrossRef]

35. Cucarella, C.; Solano, C.; Valle, J.; Amorena, B.; Lasa, I.N.I.; Penadés, J.R. Bap, a *Staphylococcus aureus* Surface Protein Involved in Biofilm Formation. *J. Bacteriol.* **2001**, *183*, 2888–2896. [CrossRef] [PubMed]

36. Ceniti, C.; Britti, D.; Santoro, A.M.L.; Musarella, R.; Ciambrone, L.; Casalinuovo, F.; Costanzo, N. Phenotypic antimicrobial resistance profile of isolates causing clinical mastitis in dairy animals. *Ital. J. Food Saf.* **2017**, *6*. [CrossRef]

37. Peixoto, R.D.M.; França, C.A.D.; Souza Júnior, A.F.D.; Veschi, J.L.A.; Costa, M.M.D. Etiologia e perfil de sensibilidade antimicrobiana dos isolados bacterianos da mastite em pequenos ruminantes e concordância de técnicas empregadas no diagnóstico. *Pesqui. Veterinária Bras.* **2010**, *30*, 735–740. [CrossRef]

38. Andrade, N.P.C.; Peixoto, R.M.; Nogueira, D.M.; Krewer, C.C.; Costa, M.M. Perfil de sensibilidade aos antimicrobianos de *Staphylococcus* spp. coagulase negativa de um rebanho leiteiro caprino em Santa Maria da Boa Vista-PE. *Med. Veterinária* **2012**, *6*, 1–6.

39. Al-Ashmawy, M.A.; Sallam, K.I.; Abd-Elghany, S.M.; Elhadidy, M.; Tamura, T. Prevalence, Molecular Characterization, and Antimicrobial Susceptibility of Methicillin-Resistant *Staphylococcus aureus* isolated from Milk and Dairy Products. *Foodborne Pathog. Dis.* **2016**, *13*, 156–162. [CrossRef] [PubMed]

40. Merz, A.; Stephan, R.; Johler, S. *Staphylococcus aureus* Isolates from Goat and Sheep Milk Seem to Be Closely Related and Differ from Isolates Detected from Bovine Milk. *Front. Microbiol.* **2016**, *7*, 319. [CrossRef] [PubMed]

41. Akkou, M.; Bentayeb, L.; Ferdji, K.; Medrouh, B.; Bachtarzi, M.-A.; Ziane, H.; Kaidi, R.; Tazir, M. Phenotypic characterization of Staphylococci causing mastitis in goats and microarray-based genotyping of *Staphylococcus aureus* isolates. *Small Rumin. Res.* **2018**, *169*, 29–33. [CrossRef]

42. Lee, J.H. Methicillin (Oxacillin)—Resistant *Staphylococcus aureus* Strains Isolated from Major Food Animals and Their Potential Transmission to Humans. *Appl. Environ. Microbiol.* **2003**, *69*, 6489–6494. [CrossRef]

43. Persson, Y.; Nyman, A.-K.; Söderquist, L.; Tomic, N.; Waller, K.P. Intramammary infections and somatic cell counts in meat and pelt producing ewes with clinically healthy udders. *Small Rumin. Res.* **2017**, *156*, 66–72. [CrossRef]

44. Queiroga, M.C.; Laranjo, M.; Duarte, E.L. Mastitis in Sheep: A disease to be Taken Seriously. In *Sheep Diseases*; Papst, M., Ed.; Nova Science Publisher, Inc: New York, NY, USA, 2019; pp. 1–91.

45. Barrero-Domínguez, B.; Luque, I.; Galán-Relaño, Á.; Vega-Pla, J.L.; Huerta, B.; Román, F.; Astorga, R.J. Antimicrobial Resistance and Distribution of *Staphylococcus* spp. Pulsotypes Isolated from Goat and Sheep Bulk Tank Milk in Southern Spain. *Foodborne Pathog. Dis.* **2019**, *16*, 723–730. [CrossRef]

46. Gosselin, V.B.; Lovstad, J.; Dufour, S.; Adkins, P.R.F.; Middleton, J.R. Use of MALDI-TOF to characterize staphylococcal intramammary infections in dairy goats. *J. Dairy Sci.* **2018**, *101*, 6262–6270. [CrossRef] [PubMed]

47. Bergonier, D.; Berthelot, X. New advances in epizootiology and control of ewe mastitis. *Livest. Prod. Sci.* **2003**, *79*, 1–16. [CrossRef]

48. Martins, K.B.; Faccioli, P.Y.; Bonesso, M.F.; Fernandes, S.; Oliveira, A.A.; Dantas, A.; Zafalon, L.F.; Cunha, M.D.L.R.S. Characteristics of resistance and virulence factors in different species of coagulase-negative staphylococci isolated from milk of healthy sheep and animals with subclinical mastitis. *J. Dairy Sci.* **2017**, *100*, 2184–2195. [CrossRef] [PubMed]

49. Park, J.Y.; Fox, L.K.; Seo, K.S.; McGuire, M.A.; Park, Y.H.; Rurangirwa, F.R.; Sischo, W.M.; Bohach, G.A. Comparison of phenotypic and genotypic methods for the species identification of coagulase-negative staphylococcal isolates from bovine intramammary infections. *Vet. Microbiol.* **2011**, *147*, 142–148. [CrossRef]

50. Becker, K.; Heilmann, C.; Peters, G. Coagulase-negative staphylococci. *Clin. Microbiol. Rev.* **2014**, *27*, 870–926. [CrossRef] [PubMed]

51. Queiroga, M.C.; Duarte, E.L.; Laranjo, M. Sheep mastitis *Staphylococcus epidermidis* biofilm effects on cell adhesion and inflammatory changes. *Small Rumin. Res.* **2018**, *168*, 6–11. [CrossRef]

52. Aher, T.; Roy, A.; Kumar, P. Molecular detection of virulence genes associated with pathogenicity of Gram positive isolates obtained from respiratory tract of apparently healthy as well as sick goats. *Vet. World* **2012**, *5*, 676. [CrossRef]

53. Saei, H.D. Coa types and antimicrobial resistance profile of *Staphylococcus aureus* isolates from cases of bovine mastitis. *Comp. Clin. Pathol.* **2012**, *21*, 301–307. [CrossRef]

54. Aarestrup, F.M.; Dangler, C.A.; Sordillo, L.M. Prevalence of coagulase gene polymorphism in *Staphylococcus aureus* isolates causing bovine mastitis. *Can. J. Vet. Res.* **1995**, *59*, 124–128.

55. Hookey, J.V.; Richardson, J.F.; Cookson, B.D. Molecular typing of *Staphylococcus aureus* based on PCR restriction fragment length polymorphism and DNA sequence analysis of the coagulase gene. *J. Clin. Microbiol.* **1998**, *36*, 1083–1089. [CrossRef] [PubMed]

56. Soltan Dallal, M.M.; Khoramizadeh, M.R.; Amiri, S.A.; Saboor Yaraghi, A.A.; Mazaheri Nezhad Fard, R. Coagulase gene polymorphism of *Staphylococcus aureus* isolates: A study on dairy food products and other foods in Tehran, Iran. *Food Sci. Hum. Wellness* **2016**, *5*, 186–190. [CrossRef]

57. Shopsin, B.; Gomez, M.; Waddington, M.; Riehman, M.; Kreiswirth, B.N. Use of Coagulase Gene (*coa*) Repeat Region Nucleotide Sequences for Typing of Methicillin-Resistant *Staphylococcus aureus* Strains. *J. Clin. Microbiol.* **2000**, *38*, 3453–3456. [CrossRef] [PubMed]

58. Tiwari, H.K.; Sapkota, D.; Gaur, A.; Mathuria, J.P.; Singh, A.; Sen, M.R. Molecular typing of clinical *Staphylococcus aureus* isolates from northern India using coagulase gene PCR-RFLP. *Southeast Asian J. Trop. Med. Public Health* **2008**, *39*, 467–473.

59. Xu, J.; Tan, X.; Zhang, X.; Xia, X.; Sun, H. The diversities of staphylococcal species, virulence and antibiotic resistance genes in the subclinical mastitis milk from a single Chinese cow herd. *Microb. Pathog.* **2015**, *88*, 29–38. [CrossRef] [PubMed]

60. Van Leeuwen, W.; Roorda, L.; Hendriks, W.; Francois, P.; Schrenzel, J. A *nuc*-deficient methicillin-resistant *Staphylococcus aureus* strain. *FEMS Immunol. Med. Microbiol.* **2008**, *54*, 157. [CrossRef] [PubMed]

61. Hirotaki, S.; Sasaki, T.; Kuwahara-Arai, K.; Hiramatsu, K. Rapid and Accurate Identification of Human-Associated Staphylococci by Use of Multiplex PCR. *J. Clin. Microbiol.* **2011**, *49*, 3627–3631. [CrossRef]

62. Silva, W.P.D.; Silva, J.A.; Macedo, M.R.P.D.; Araújo, M.R.D.; Mata, M.M.; Gandra, E.A. Identification of *Staphylococcus aureus, S. intermedius* and *S. hyicus* by PCR amplification of *coa* and *nuc* genes. *Braz. J. Microbiol.* **2003**, *34*, 125–127. [CrossRef]

63. Cucarella, C.; Tormo, M.A.; Úbeda, C.; Trotonda, M.P.; Monzón, M.; Peris, C.; Amorena, B.; Lasa, I.; Penadés, J.R. Role of Biofilm-Associated Protein Bap in the Pathogenesis of Bovine *Staphylococcus aureus*. *Infect. Immun.* **2004**, *72*, 2177–2185. [CrossRef]

64. Vautor, E.; Magnone, V.; Rios, G.; Le Brigand, K.; Bergonier, D.; Lina, G.; Meugnier, H.; Barbry, P.; Thiéry, R.; Pépin, M. Genetic differences among *Staphylococcus aureus* isolates from dairy ruminant species: A single-dye DNA microarray approach. *Vet. Microbiol.* **2009**, *133*, 105–114. [CrossRef]

65. Martins, K.B.; Faccioli-Martins, P.Y.; Riboli, D.F.; Pereira, V.C.; Fernandes, S.; Oliveira, A.A.; Dantas, A.; Zafalon, L.F.; da Cunha, M.E.L. Clonal profile, virulence and resistance of *Staphylococcus aureus* isolated from sheep milk. *Braz. J. Microbiol.* **2015**, *46*, 535–543. [CrossRef]

66. Tormo, M.Á.; Knecht, E.; Götz, F.; Lasa, I.; Penadés, J.R. Bap-dependent biofilm formation by pathogenic species of *Staphylococcus*: Evidence of horizontal gene transfer? *Microbiology* **2005**, *151*, 2465–2475. [CrossRef] [PubMed]

67. Szweda, P.; Schielmann, M.; Milewski, S.; Frankowska, A.; Jakubczak, A. Biofilm production and presence of ica and bap genes in *Staphylococcus aureus* strains isolated from cows with mastitis in the eastern Poland. *Pol. J. Microbiol.* **2012**, *61*, 65–69. [CrossRef] [PubMed]

68. Marques, V.F.; Motta, C.C.D.; Soares, B.D.S.; Melo, D.A.D.; Coelho, S.D.M.D.O.; Coelho, I.D.S.; Barbosa, H.S.; Souza, M.M.S.D. Biofilm production and beta-lactamic resistance in Brazilian *Staphylococcus aureus* isolates from bovine mastitis. *Braz. J. Microbiol.* **2017**, *48*, 118–124. [CrossRef] [PubMed]

69. Khoramian, B.; Jabalameli, F.; Niasari-Naslaji, A.; Taherikalani, M.; Emaneini, M. Comparison of virulence factors and biofilm formation among *Staphylococcus aureus* strains isolated from human and bovine infections. *Microb. Pathog.* **2015**, *88*, 73–77. [CrossRef]

70. Vitale, M.; Galluzzo, P.; Buffa, P.G.; Carlino, E.; Spezia, O.; Alduina, R. Comparison of Antibiotic Resistance Profile and Biofilm Production of *Staphylococcus aureus* Isolates Derived from Human Specimens and Animal-Derived Samples. *Antibiotics* **2019**, *8*, 97. [CrossRef]

71. Zuniga, E.; Melville, P.A.; Saidenberg, A.B.S.; Laes, M.A.; Gonsales, F.F.; Salaberry, S.R.S.; Gregori, F.; Brandão, P.E.; dos Santos, F.G.B.; Lincopan, N.E.; et al. Occurrence of genes coding for MSCRAMM and biofilm-associated protein Bap in *Staphylococcus* spp. isolated from bovine subclinical mastitis and relationship with somatic cell counts. *Microb. Pathog.* **2015**, *89*, 1–6. [CrossRef]

72. Notcovich, S.; DeNicolo, G.; Flint, S.; Williamson, N.; Gedye, K.; Grinberg, A.; Lopez-Villalobos, N. Biofilm-Forming Potential of *Staphylococcus aureus* Isolated from Bovine Mastitis in New Zealand. *Vet. Sci.* **2018**, *5*, 8. [CrossRef]

73. Magiorakos, A.P.; Srinivasan, A.; Carey, R.B.; Carmeli, Y.; Falagas, M.E.; Giske, C.G.; Harbarth, S.; Hindler, J.F.; Kahlmeter, G.; Olsson-Liljequist, B.; et al. Multidrug-resistant, extensively drug-resistant and pandrug-resistant bacteria: An international expert proposal for interim standard definitions for acquired resistance. *Clin. Microbiol. Infect.* **2012**, *18*, 268–281. [CrossRef]

74. França, C.A.; Peixoto, R.M.; Cavalcante, M.B.; Melo, N.F.; Oliveira, C.J.B.; Veschi, J.A.; Mota, R.A.; Costa, M.M. Antimicrobial resistance of *Staphylococcus* spp. from small ruminant mastitis in Brazil. *Pesqui. Veterinária Bras.* **2012**, *32*, 747–753. [CrossRef]

75. Taponen, S.; Pyorala, S. Coagulase-negative staphylococci as cause of bovine mastitis—Not so different from *Staphylococcus aureus*? *Vet. Microbiol.* **2009**, *134*, 29–36. [CrossRef] [PubMed]

76. Vasileiou, N.G.C.; Sarrou, S.; Papagiannitsis, C.; Chatzopoulos, D.C.; Malli, E.; Mavrogianni, V.S.; Petinaki, E.; Fthenakis, G.C. Antimicrobial Agent Susceptibility and Typing of Staphylococcal Isolates from Subclinical Mastitis in Ewes. *Microb. Drug Resist.* **2019**, *25*, 1099–1110. [CrossRef] [PubMed]

77. Queiroga, M.C.; Andrade, N.; Laranjo, M. Antimicrobial Action of Propolis Extracts Against Staphylococci. In *Understanding Microbial Pathogens: Current Knowledge and Educational Ideas on Antimicrobial Research*; Torres-Hergueta, E., Méndez-Vilas, A., Eds.; Formatex Research Center: Badajoz, Spain, 2018; pp. 28–35.

78. CLSI. *M100S-Performance Standards for Antimicrobial Susceptibility Testing-Approved Standard*, 26th ed.; Clinical and Laboratory Standards Institute: Wayne, PA, USA, 2016.

79. Reading, C.; Cole, M. Clavulanic acid: A beta-lactamase-inhiting beta-lactam from *Streptomyces clavuligerus*. *Antimicrob. Agents Chemother.* **1977**, *11*, 852–857. [CrossRef] [PubMed]

80. Barbier, F.; Ruppé, E.; Hernandez, D.; Lebeaux, D.; Francois, P.; Felix, B.; Desprez, A.; Maiga, A.; Woerther, P.L.; Gaillard, K.; et al. Methicillin-Resistant Coagulase-Negative Staphylococci in the Community: High Homology of SCCmec IVa between *Staphylococcus epidermidis* and Major Clones of Methicillin-Resistant *Staphylococcus aureus*. *J. Infect. Dis.* **2010**, *202*, 270–281. [CrossRef] [PubMed]

81. Yamada, K.; Namikawa, H.; Fujimoto, H.; Nakaie, K.; Takizawa, E.; Okada, Y.; Fujita, A.; Kawaguchi, H.; Nakamura, Y.; Abe, J.; et al. Clinical Characteristics of Methicillin-resistant Coagulase-negative Staphylococcal Bacteremia in a Tertiary Hospital. *Intern. Med.* **2017**, *56*, 781–785. [CrossRef] [PubMed]

82. El-Razik, K.A.A.; Arafa, A.A.; Hedia, R.H.; Ibrahim, E.S. Tetracycline resistance phenotypes and genotypes of coagulase-negative staphylococcal isolates from bubaline mastitis in Egypt. *Vet. World* **2017**, *10*, 702–710. [CrossRef]

83. El Feghaly, R.E.; Stamm, J.E.; Fritz, S.A.; Burnham, C.-A.D. Presence of the *blaZ* beta-lactamase gene in isolates of *Staphylococcus aureus* that appear penicillin susceptible by conventional phenotypic methods. *Diagn. Microbiol. Infect. Dis.* **2012**, *74*, 388–393. [CrossRef]

84. Paterson, G.K.; Harrison, E.M.; Holmes, M.A. The emergence of *mecC* methicillin-resistant *Staphylococcus aureus*. *Trends Microbiol.* **2014**, *22*, 42–47. [CrossRef]

85. Xu, Z.; Shi, L.; Alam, M.J.; Li, L.; Yamasaki, S. Integron-bearing methicillin-resistant coagulase-negative staphylococci in South China, 2001–2004. *FEMS Microbiol. Lett.* **2008**, *278*, 223–230. [CrossRef]

86. Becker, K.; van Alen, S.; Idelevich, E.A.; Schleimer, N.; Seggewiß, J.; Mellmann, A.; Kaspar, U.; Peters, G. Plasmid-Encoded Transferable *mecB*-Mediated Methicillin Resistance in *Staphylococcus aureus*. *Emerg. Infect. Dis.* **2018**, *24*, 242–248. [CrossRef]

87. Akpaka, P.E.; Roberts, R.; Monecke, S. Molecular characterization of antimicrobial resistance genes against Staphylococcus aureus isolates from Trinidad and Tobago. *J. Infect. Public Health* **2017**, *10*, 316–323. [CrossRef]

88. Yılmaz, E.Ş.; Aslantaş, Ö. Antimicrobial resistance and underlying mechanisms in Staphylococcus aureus isolates. *Asian Pac. J. Trop. Med.* **2017**, *10*, 1059–1064. [CrossRef] [PubMed]

89. Markey, B.; Leonard, F.; Archambault, M.; Cullinane, A.; Maguire, D. *Clinical Veterinary Microbiology*, 2nd ed.; Mosby Ltd.: London, UK, 2013.

90. Laranjo, M.; Machado, J.; Young, J.P.W.; Oliveira, S. High diversity of chickpea *Mesorhizobium* species isolated in a Portuguese agricultural region. *FEMS Microbiol. Ecol.* **2004**, *48*, 101–107. [CrossRef] [PubMed]

91. Merino, N.; Toledo-Arana, A.; Vergara-Irigaray, M.; Valle, J.; Solano, C.; Calvo, E.; Lopez, J.A.; Foster, T.J.; Penadés, J.R.; Lasa, I. Protein A-mediated multicellular behavior in Staphylococcus aureus. *J. Bacteriol.* **2009**, *191*, 832–843. [CrossRef] [PubMed]

92. CLSI. *M02-A11-Performance Standards for Antimicrobial Disk Susceptibility Tests-Approved Standard-Eleventh Edition*; Clinical and Laboratory Standards Institute (CLSI): Wayne, PA, USA, 2012.

93. EUCAST. *European Committee on Antimicrobial Susceptibility Testing-Breakpoint Tables for Interpretation of MICs and Zone Diameters*; Version 10.0; UCAST: Växjö, Sweden, 2020.

94. Rivas, R.; Vizcaino, N.; Buey, R.M.; Mateos, P.F.; Martinez-Molina, E.; Velazquez, E. An effective, rapid and simple method for total RNA extraction from bacteria and yeast. *J. Microbiol. Methods* **2001**, *47*, 59–63. [CrossRef]

95. Brakstad, O.D.D.G.; Aasbakk, K.; Maeland, J.A. Detection of *Staphylococcus aureus* by Polymerase Chain Reaction Amplification of the *nuc* Gene. *J. Clin. Microbiol.* **1992**, *30*, 1654–1660. [CrossRef] [PubMed]
96. Vasudevan, P.; Nair, M.K.; Annamalai, T.; Venkitanarayanan, K.S. Phenotypic and genotypic characterization of bovine mastitis isolates of *Staphylococcus aureus* for biofilm formation. *Vet. Microbiol.* **2003**, *92*, 179–185. [CrossRef]
97. Sawant, A.; Gillespie, B.; Oliver, S. Antimicrobial susceptibility of coagulase-negative *Staphylococcus* species isolated from bovine milk. *Vet. Microbiol.* **2009**, *134*, 73–81. [CrossRef]
98. Strommenger, B.; Kettlitz, C.; Werner, G. Multiplex PCR assay for simultaneous detection of nine clinically relevant antibiotic resistance genes in *Staphylococcus aureus*. *J. Clin. Microbiol.* **2003**, *41*, 4089–4094. [CrossRef]
99. Stegger, M.; Andersen, P.S.; Kearns, A.; Pichon, B.; Holmes, M.A.; Edwards, G.; Laurent, F.; Teale, C.; Skov, R.; Larsen, A.R. Rapid detection, differentiation and typing of methicillin-resistant *Staphylococcus aureus* harbouring either *mecA* or the new *mecA* homologue *mecALGA251*. *Clin. Microbiol. Infect.* **2012**, *18*, 395–400. [CrossRef]
100. Ferreira, A.M.; Martins, K.B.; Silva, V.R.D.; Mondelli, A.L.; Cunha, M.D.L.R.D.S.D. Correlation of phenotypic tests with the presence of the *blaZ* gene for detection of beta-lactamase. *Braz. J. Microbiol.* **2017**, *48*, 159–166. [CrossRef]
101. Horstkotte, M.A.; Knobloch, J.K.M.; Rohde, H.; Mack, D. Rapid Detection of Methicillin Resistance in Coagulase-Negative Staphylococci by a Penicillin-Binding Protein 2a-Specific Latex Agglutination Test. *J. Clin. Microbiol.* **2001**, *39*, 3700–3702. [CrossRef] [PubMed]
102. Coelho, S.D.M.D.O.; Moraes, R.A.M.; Soares, L.D.C.; Pereira, I.A.; Gomes, L.P.; Souza, M.M.S.D. Mapeamento do perfil de resistência e detecção do gene *mecA* em *Staphylococcus aureus* e *Staphylococcus intermedius* oxacilina-resistentes isolados de espécies humanas e animais. *Ciência Rural* **2007**, *37*, 195–200. [CrossRef]

 *antibiotics*

Article

Virulence, Antimicrobial Resistance and Biofilm Production of *Escherichia coli* Isolates from Healthy Broiler Chickens in Western Algeria

Qada Benameur [1], Teresa Gervasi [2,*], Filippo Giarratana [3], Maria Vitale [4], Davide Anzà [4], Erminia La Camera [5], Antonia Nostro [5], Nicola Cicero [2] and Andreana Marino [5]

1   Nursing Department, Faculty of Nature and Life Sciences, University of Mostaganem, Mostaganem 27000, Algeria; qada.benameur@univ-mosta.dz

2   Department of Biomedical and Dental Sciences and Morphofunctional Imaging, University of Messina, 98100 Messina, Italy; nicola.cicero@unime.it

3   Department of Veterinary Sciences, University of Messina, 98100 Messina, Italy; filippo.giarratana@unime.it

4   Istituto Zooprofilattico Sperimentale della Sicilia "Adelmo Mirri", 90141 Palermo, Italy; maria.vitale@izssicilia.it (M.V.); davide.anza@gmail.com (D.A.)

5   Department of Chemical, Biological, Pharmaceutical and Environmental Sciences, University of Messina, 98100 Messina, Italy; erminia.lacamera@unime.it (E.L.C.); antonia.nostro@unime.it (A.N.); andreana.marino@unime.it (A.M.)

*   Correspondence: tgervasi@unime.it; Tel.: +39-090-676-2870

Citation: Benameur, Q.; Gervasi, T.; Giarratana, F.; Vitale, M.; Anzà, D.; La Camera, E.; Nostro, A.; Cicero, N.; Marino, A. Virulence, Antimicrobial Resistance and Biofilm Production of *Escherichia coli* Isolates from Healthy Broiler Chickens in Western Algeria. *Antibiotics* **2021**, *10*, 1157. https:// doi.org/10.3390/antibiotics10101157

Academic Editors: Manuela Oliveira and Elisabete Silva

Received: 26 August 2021
Accepted: 22 September 2021
Published: 24 September 2021

Publisher's Note: MDPI stays neutral with regard to jurisdictional claims in published maps and institutional affiliations.

Abstract: The aim of this study was to assess the virulence, antimicrobial resistance and biofilm production of *Escherichia coli* strains isolated from healthy broiler chickens in Western Algeria. *E. coli* strains (n = 18) were identified by matrix-assisted laser desorption–ionization time-of-flight mass spectrometry. Susceptibility to 10 antibiotics was determined by standard methods. Virulence and extended-spectrum β-lactamase (ESBL) genes were detected by PCR. The biofilm production was evaluated by microplate assay. All the isolates were negative for the major virulence/toxin genes tested (*rfbE*, *fliC*, *eaeA*, *stx1*), except one was *stx2*-positive. However, all were resistant to at least three antibiotics. Ten strains were ESBL-positive. Seven carried the β-lactamase *bla*$_{TEM}$ gene only and two co-harbored *bla*$_{TEM}$ and *bla*$_{CTX-M-1}$ genes. One carried the *bla*$_{SHV}$ gene. Among the seven strains harboring *bla*$_{TEM}$ only, six had putative enteroaggregative genes. Two contained *irp2*, two contained both *irp2* and *astA*, one contained *astA* and another contained *aggR*, *astA* and *irp2* genes. All isolates carrying ESBL genes were non-biofilm producers, except one weak producer. The ESBL-negative isolates were moderate biofilm producers and, among them, two harbored *astA*, two *irp2*, and one *aggR*, *astA* and *irp2* genes. This study highlights the spread of antimicrobial-resistant *E. coli* strains from healthy broiler chickens in Western Algeria.

Keywords: virulence genes; antimicrobial resistance; extended-spectrum β-lactamases; biofilm formation

1. Introduction

One of the most important global challenges to public health is represented by foodborne illnesses. Healthy food-producing animals can be vectors for a wide range of commensal and pathogenic bacteria, as well as *Escherichia coli*. This microorganism can contaminate the food chain at each step, from the slaughterhouses to the food processing phases [1–4]. To date, although several *E. coli* strains are commensals, which colonize the gastrointestinal tract of humans and warm-blooded animals, and are not often disease-causing, *E. coli* represents one of the most frequent causes of several common infections in humans and animals [5]. *E. coli* clones acquiring specific virulence factors (VFs), including adhesins, toxins, invasins, etc., can cause intestinal and extra-intestinal diseases such as enteric/diarrheal disease, urinary tract infections (UTIs) and sepsis/meningitis in human hosts [6–10].

VFs are generally carried on phages, plasmids or pathogenicity islands (PAIs) [11] and, among microbial strains, can be vastly interchanged via horizontal transfer [12]. Given the presence of definite virulence genes, *E. coli* strains can be classified as pathogens [13,14], in particular as zoonotic intestinal pathogenic *E. coli* pathotypes (IPEC) or extraintestinal pathogenic *E. coli* pathotypes (ExPEC) based on the type of VFs present and the host's clinical symptoms [15]. The specific "pathotypes" are grouped into enteropathogenic *E. coli* (EPEC), enterohaemorrhagic *E. coli* (EHEC), enterotoxigenic *E. coli* (ETEC), enteroaggregative *E. coli* (EAEC), enteroinvasive *E. coli* (EIEC) and diffusely adherent *E. coli* (DAEC), and can cause intestinal diseases [8,16].

Moreover, the emergence of antibiotic resistance in food pathogens represents further complications.

The wide use of antibiotics, in both animals and humans, is responsible for an increased antibiotic resistance not only in pathogenic bacteria, but also in the endogenous microflora. Resistant animal bacteria can be transmitted to humans through several routes, such as direct contact with the animal or its manure, and through contact with or the consumption of uncooked meat [17–21]. Given the development of combined resistance to multiple antibiotics such as the β-lactam group, including cephalosporins and carbapenems, over the last few years, the chemotherapeutic choices for enterobacteria are becoming strictly limited [22]. Resistance to cephalosporin is the result of the production of one or more types of β-lactamases, the so called extended spectrum-β-lactamases (ESBLs) [23]. ESBLs are categorized into several classes, among which the most important include Temoniera (TEM), sulfhydryl variable (SHV) and cefotaximase (CTX-M) types [24,25]. Thus, nowadays ESBL-producing Gram-negative bacteria represent a growing concern and an important challenge for chemotherapy [26]. In addition, another issue is represented by the fact that, in food factory environments, some biofilm-forming bacteria are human pathogens. Biofilms are ecosystems made up of one or more bacterial species submerged in an extracellular matrix, whose composition varies according to the colonizing species and the food manufacturing environment [27–29].

The zoonotic potential of *E. coli* from chicken food products is important for public health purposes [30,31]. Meat harbors different bacteria as inherent contamination and is further contaminated during handling, improper dressing, cleaning, unsanitary conditions and unhygienic practices during its commercialization [32]. Considering the factors described, the objective of this preliminary study was to examine virulence and antimicrobial resistance (AMR) gene profiles, and the ability of biofilm formation of *E. coli* strains isolated from healthy broiler chickens in Western Algeria. The Algerian poultry industry, consisting of 20,000 farmers, every year yields an average of 340,000 tons of white meat and over 4.8 billion eggs. The present Algerian poultry industry structure is the result of government development policies, which were initiated in the 1980s [33].

2. Results

A total of 18 presumptive *E. coli* strains were isolated from 32 fecal samples, collected from different broiler chicken farms situated in five geographic areas of Western Algeria: Mostaganem (n = 8, 25.0%), Oran (n = 6, 18.75%), Mascara (n = 6, 18.75%), Relizane (n = 6, 18.75%) and Tiaret (n = 6, 18.75%). MALDI-TOF-MS analysis confirmed the identification of all the 18 presumptive *E. coli* strains (Table 1). All the isolates were negative for the major virulence/toxin genes tested, including shiga-like toxin 1 (*stx1*), O157:H7 serotype (*rfbE*), flagellar gene (*fliC*) and attaching and effacing gene (*eaeA*), except for one *E. coli* strain positive for the shiga-like toxin 2 (*stx2*) gene (Table 1), coming from 1/7 broiler houses located in the Mostaganem area.

Table 1. Characteristics of *E. coli* isolates.

Strains	Algerian Area	Virulence Gene [1]	MALDI-TOF Mean Value	*bla* Gene [2]	AMR [3]
S13/15	Oran	*astA*	90.0%	None	NA, CIP, AML, SXT, TE, N
S14/15	Oran	*irp2*	87.4%	CTX-M-1, TEM	NA, CIP, AUG, SXT, TE, N, CTX
S2/15	Oran	*irp2*	85.1%	TEM	NA, CIP, AML, AUG, SXT, TE, N, CTX
S4/15	Mostaganem	*stx2*	87.5%	None	NA, AML, AUG, SXT, TE, N
S19/15	Mostaganem	None	93.0%	CTX-M-1, TEM	NA, CIP, AML, AUG, SXT, TE, C, N, CTX
S12/15	Mostaganem	*aggR, astA, irp2*	85.8%	TEM	NA, CIP, AML, AUG, SXT, TE, N, CTX
S25a/16	Mostaganem	*astA*	88.7%	None	NA, CIP, AUG, TE, N
S1/16	Mostaganem	*astA, irp2*	97.7%	TEM	NA, CIP, AML, AUG, SXT, TE, N, CTX
S22/15	Mostaganem	*astA, irp2*	89.4%	TEM	NA, CIP, AML, AUG, SXT, TE, N, CTX
S16/15	Mostaganem	*irp2*	92.6%	None	NA, CIP, AML, AUG, SXT, TE, N
S34/16	Relizane	*irp2*	93.3%	None	NA, CIP, N
S31/16	Relizane	None	91.2%	None	NA, CIP, TE, N
S33/16	Relizane	*aggR, astA, irp2*	90.3%	None	NA, CIP, AML, AUG, SXT, TE, C, N
S47/16	Tiaret	*irp2*	95.2%	TEM	NA, CIP, AML, AUG, SXT, TE, N, CTX
S6/15	Tiaret	None	96.5%	None	NA, CIP, AML, AUG, TE, N
S48a/16	Tiaret	*astA*	93.4%	TEM	NA, CIP, AML, AUG, SXT, TE, C, N, CTX
S19a/16	Mascara	None	92.7%	TEM	NA, CIP, AML, AUG, SXT, TE, N, CTX
S61a/16	Mascara	None	95.0%	SHV	NA, AUG, SXT, TE, N
E. coli ATCC 259222			99.9%		AML, AUG

[1] *astA*, heat-stable enterotoxin-1; *irp2*, iron regulatory protein 2; *stx2*, shiga-like toxin 2; *aggR*, transcription factor; [2] TEM, temoniera; CTX-M-1, cefotaximases; SHV, sulfhydryl variable; [3] AMR, antimicrobial resistance; NA, nalidixic acid; N, neomycin; TE, tetracycline; CIP, ciprofloxacin; AUG, amoxicillin–clavulanic acid; STX, trimethoprim–sulfamethoxazole; AML, amoxicillin; CTX, cefotaxime; C, chloramphenicol.

However, in contrast to the low percentage of virulence genes detected, all strains were shown to be resistant to at least three antibiotics most frequently used in poultry (Table 1). They were resistant to nalidixic acid (NA) (100%), neomycin (N) (100%), tetracycline (TE) (94%), ciprofloxacin (CIP) (89%), amoxicillin–clavulanic acid (AUG) (83%), trimethoprim–sulfamethoxazole (SXT) (78%), amoxicillin (AML) (72%), cefotaxime (CTX) (50%) and chloramphenicol (C) (17%). Among the strains, 10 were phenotypically confirmed to be ESBL-positive isolates. Genotypic analyses revealed that nine strains (CTX-resistant 50%) carried the *bla*$_{TEM}$ gene and one harbored the *bla*$_{SHV}$ gene (5.55%). Among the *bla*$_{TEM}$-producing *E. coli* isolates, two co-harbored the *bla*$_{CTX-M-1}$ gene (11%) (Table 2). The distribution of the percentages of ESBL isolates and the geographical area visited was 67% in Oran, 57% in Mostaganem, 67% in Tiaret, 50% in Mascara and none in Relizane.

Furthermore, an association between biofilm production and the presence of enteroaggregative genes was evaluated.

Table 2. Enteroaggregative and ESBL genes and biofilm production of *E. coli* isolates.

Strains	*aggR*	*irp2*	*astA*	TEM	CTX-M-1	SHV	SBF	Biofilm Grade
S13/15			+				0.81	M
S14/15		+		+	+		0.16	N
S2/15		+		+			0.26	N
S4/15 *stx2*							0.76	M
S19/15				+	+		0.20	N
S12/15	+	+	+	+			0.18	N
S25a/16			+				0.62	W
S1/16		+	+	+			0.19	N
S22/15		+	+	+			0.29	N
S16/15		+					0.45	W
S34/16		+					0.52	W
S31/16							0.84	M
S33/16	+	+	+				0.65	W
S47/16		+		+			0.39	W
S6/15							0.59	W
S48a/16			+	+			0.22	N
S19a/16				+			0.24	N
S61a/16						+	0.27	N
E. coli ATCC 25922							0.76	M

aggR, transcription factor; *irp2*, iron regulatory protein 2; *astA*, heat-stable enterotoxin-1; TEM, temoniera; CTX-M-1, cefotaximases; SHV, sulfhydryl variable; SBF, specific biofilm formation; + gene presence; M, moderate (SBF $\geq$ 0.70–1.09); N, negative (SBF $<$ 0.35); W, weak (SBF $\geq$ 0.35–0.69).

Enteroaggregative genes were detected in the ESBL-producing strains (Table 1). Among the seven strains harboring only $bla_{\text{TEM}-1}$, two strains contained iron regulatory protein 2 (*irp2*) (28.5%), two both *irp2* and the heat-stable enterotoxin-1 (*astA*) (28.5%), one *astA* (14%) and another the transcription factor (*aggR*), *astA* and *irp2* (14%) genes. Two strains contained $bla_{\text{TEM}}/bla_{\text{CTXM}-1}$, and one had the *irp2* (50%) gene. Among all ESBL-producing strains, only one isolate, harboring bla_{TEM} and *irp2* genes, was a weak biofilm producer (14%) (Table 2). The remaining strains (86%) were regarded as non-biofilm producers (specific biofilm formation (SBF): 0.16–0.29).

Among the eight non-ESBL-producing strains, five (62.5%) harbored putative enteroaggregative genes: *astA* (n = 2, 25%), *irp2* (n = 2, 25%) and *astA–irp2–aggR* (n = 1, 12.5%). Moreover, one isolate (12.5%) expressed the *stx2* gene (Table 1). The non-ESBL-producing isolates were more likely to produce biofilm than ESBL-producing strains ($p \geq 0.001$; r = 0.85). Among the non-ESBL-producing isolates, three strains were classified as moderate biofilm producers (Table 2). One harbored *astA* (12.5%), another *stx2* (12.5%) and yet another contained no virulence gene (12.5%), with SBF: 0.81, SBF: 0.76 and SBF: 0.84, respectively. The remaining strains (62.5%) were regarded as weak biofilm producers (SBF: 0.39–0.65). *E. coli* ATCC 25922 was a moderate biofilm producer (SBF: 0.81).

3. Discussion

The majority of *E. coli* strains are commensals inhabiting the intestinal tract of humans and warm-blooded animals and rarely cause diseases, unless they acquire VFs carried by mobile genetic elements such as bacteriophages, pathogenicity islands and plasmids [34]. Additionally, *E. coli* can form a reservoir of AMR genes that may be transferred among different bacterial species, including pathogenic bacteria for both humans and animals.

In this study, the *E. coli* strains, isolated from fecal samples of apparently healthy chickens, showed a low percentage of virulence genes, which are characteristic of shiga toxin-producing *E. coli* (STEC O157:H7) (*rfbE, fliC, eaeA* and *stx1*). Indeed, except for one *E. coli* strain, which was positive for the *stx2* gene detected at one Mostaganem farm, all the isolates were negative for the major genes encoding VFs. This finding is in accordance with previous Algerian studies describing a low prevalence of *stx* genes in *E. coli* isolates from poultry origin, i.e., a recent Algerian study showed the presence of *stx2* in only one *E. coli* isolate from broiler chickens, which had just died [35]. Another study conducted in the north of Algeria revealed the total absence of the *stx2* gene and the presence of the *stx1* gene in only two *E. coli* strains isolated from diarrheic hens and chickens [36]. Our results are also in agreement with several previous studies conducted in other countries, which concluded that the prevalence of STEC O157:H7 in broiler chickens is relatively low compared with other animal species [37–40].

However, in contrast to the low percentage of STEC virulence genes detected, all isolated strains were shown to be resistant to at least three antibiotics most frequently administered to poultry. Antimicrobial agents are being used in many countries in veterinary practice for the treatment of disease, disease prevention and growth promotion [41]. However, the indiscriminate use of antimicrobials can result in bacterial selection pressure of the intestinal microbiota of animals [19,42,43].

The high levels of resistance of the isolated strains against more than three antibiotics were not surprising given the uncontrolled use of these antibiotics in poultry in Algeria and their use without prior antimicrobial susceptibility tests. It must also be noted that the lack of legislative restrictions on antibiotic use in the poultry industry could also lead to a build-up of antibiotic resistance, i.e., they are still used in poultry feeds at sub-therapeutic dosages for growth promotion purposes (to reduce bird mortality and improve production performance). In contrast, this practice is banned in many countries, including those of the European Union, to avoid AMR diffusion in pathogenic bacteria in food-producing animals [44]. The high level of resistance recorded in this study for CTX (50%) is troubling as third-generation cephalosporins (ceftiofur) are not used in Algerian poultry production. These results are in agreement with those reported in other studies [45,46], which highlighted the emergence and persistence of ESBL-producing *E. coli* in the poultry production pyramid in many countries despite the absence of third-generation cephalosporin usage. This might be linked to the abuse and misuse of other antimicrobials (i.e., aminoglycosides, β-lactams, quinolones, macrolides, nitrofurans, phenicols, polypeptides, sulphonamides and tetracyclines) in broiler breeding or to the selection of ESBL-producing *E. coli* in broiler breeders and their vertical transmission in the poultry production pyramid [47–50]. The high levels of ESBL-producing *E. coli* isolates in Mostaganem, Oran, Mascara, Relizane and Tiaret could be explained by their horizontal transmission in broiler farms and hatcheries, as previously suggested [51], during broiler chicken transfer and likewise through national trade to several regions of the country. In addition, encoding cephalosporin resistance genes are generally placed on self-transmissible plasmids [52], which can be promiscuous and are capable of circulating among a wide variety of hosts. Despite the fact that third-generation cephalosporins are not used in Algerian poultry production, several studies highlighted their colonization in broiler chickens in the last few years [53–55]. The genetic background for cephalosporin resistance was diverse in those studies. Benameur et al. [54] reported the presence of the $bla_{\text{CTX-M}-1}$ gene and Meguenni et al. [55] showed the presence of $bla_{\text{CTX-M}-1}$ and $bla_{\text{CTX-M}-15}$. Furthermore, Belmahdi et al. [53] detected the presence of $bla_{\text{CTX-M}-1}$, $bla_{\text{SHV}-12}$ and $bla_{\text{TEM}-1}$.

However, in our study, the most prevalent group was bla_{TEM} followed by bla_{TEM} and $bla_{\text{CTX-M}-1}$ gene combinations and bla_{SHV}, like the findings in a study in Turkey that demonstrated bla_{TEM} as the most frequent gene, followed by $bla_{\text{CTX-M}}$ and bla_{SHV} [56].

In many other studies, multiple occurrences of the genes were also common [57], given that these genes frequently exist in large plasmids [58]. SHV and TEM were the main types of ESBL until 2000, while, in recent decades, CTX-M enzymes took their place [59].

All genes encoding resistance to macrolides, quinolones, tetracyclines, aminoglycosides, trimethoprim, chloramphenicol and sulfonamides have been associated with plasmids containing the $bla_{\text{CTX-M}}$ type [60]. The association of antibiotics, and the coexistence of $bla_{\text{CTX-M}}$ genes with bla_{TEM} or other resistance determinants, could contribute to the spread of CTX-M enzymes. Nowadays, enzymes of the CTX-M-1 group are frequently identified in North Africa [61].

This issue is further worsened by the formation of biofilm, which promotes an additional bacterial tolerance or resistance to antimicrobial agents [29,62] and represents an advantage in the survival against host defense factors, antibiotics, physical and chemical stress as well as disinfectants [63,64].

In this study, the ability of biofilm formation was found to be significantly higher in negative ESBL strains of *E. coli* than in strains carrying the *bla*TEM gene. However, despite the small number of strains used in this study, the results align with those of other authors who demonstrate that the expression of the *bla*TEM gene can negatively impact biofilm formation in *E. coli* [65].

The production of biofilm is also regulated by putative enteroaggregative genes such as the transcription activator known as *"aggR"*, the master regulator of EAEC virulence, which controls the expression of adherence factors, and several other genes including yersiniabactin operon (*irp2*) and EAST1 toxin (*astA*) [61].

However, according to other authors, no correlation was reported between *aggR* alone or in association with *irp2* and *astA* and biofilm formation in producing isolates, indicating that there are additional factors involved in biofilm production in EAEC [9,66,67].

4. Materials and Methods

4.1. Study Area and Sampling

A total of 16 broiler farms were randomly selected to carry out this study. All the farms were located within five geographic areas of Western Algeria, namely Mostaganem, Oran, Mascara, Relizane and Tiaret, representing the major broiler poultry producing sites in Algeria. Each broiler farm comprised several houses. Two poultry houses were sampled from each farm and one sample per house was collected. The poultry houses were chosen by considering their capacities (at least 3000 birds per house). All the farms included in this study were under control by official veterinary services. Broilers were commonly kept for a short period, which is generally less than two months. All broiler farms were visited once and sampling was carried out a few days before submission of the birds to slaughter. Fresh (still soft and warm) poultry feces was sampled from the poultry houses and transported to the laboratory for isolation of *E. coli*. All sampled broiler flocks were apparently healthy on the day of sampling.

4.2. Escherichia coli Isolation

Between March and September 2020, a total of 32 fecal samples, collected from different broiler chicken farms situated in five geographic areas of Western Algeria (Mostaganem, Oran, Mascara, Relizane, Tiaret), were analyzed in this study. To isolate *E. coli*, one gram of fecal specimens was mixed with 9 mL of buffered peptone water and incubated for 18 h at 37 °C. A drop was then streaked on MacConkey agar (MAC, Oxoid, Hampshire, UK) plates and incubated for 18 h at 37 °C. *E. coli* ATCC 25922 and ATCC 10536 (American Type Culture Collection, Rockville, MD, USA) were used as reference strains. Single colonies were stored in glycerol at −80 °C until further testing.

4.3. Identification of Colonies by MALDI-TOF-MS

The presumptive *E. coli* colonies were identified by matrix-assisted laser desorption–ionization time-of-flight mass spectrometry (MALDI-TOF-MS). Briefly, strains were cultured on tryptic soy agar (TSA; Oxoid, Hampshire, UK) supplemented with 5% of sheep blood and incubated at 37 °C for 24 h.

A single bacterial colony was deposited on FlexiMass MALDI target plates, with 48-well sample spots (bioMérieux, Firenze, Italy), followed by the addition of 1 μL of matrix of alpha-cyano-4-hydroxycinnamic acid matrix in 50% acetonitrile and 2.5% trifluoroacetic acid (Vitek MS-CHCA, bioMérieux, Firenze, Italy).

E. coli ATCC 8739, used as a calibrator and internal ID control, grown on TSA, which was supplemented with 5% of sheep blood (according to the constructor procedure) and incubated at 37 °C for 24 h, was inoculated on the calibration spots as well as the test strains.

The prepared plate, after the complete crystallization of the microbial matrix complex, was inserted in a Vitek MS Axima Assurance linear mass spectrometer (bioMérieux, Firenze, Italy) set with a laser frequency of 50 Hz, an acceleration voltage of 20 kV, an extraction delay time of 200 ns and mass spectra from 2000 to 20,000 Da. Every single strain was analyzed three times in three separate runs at different times.

The obtained mass spectra for each microorganism were analyzed by SARAMIS software (Spectral ARchive And Microbial Identification System—Database version 4.10—Software year 2010, bioMérieux, Firenze, Italy) by comparing them with the database bacteria reference spectra. The result of this comparison, calculated by the software algorithm, is a percentage probability (confidence level) that represents the similarity (presence or absence of specific peaks) among the obtained spectra and the reference spectra.

A perfect match reported as "excellent ID" corresponded to a percentage probability of identification (confidence level) of 99.9%, a "good ID" from >60% to 99.8%, while for <60% "no identification" (no ID) was given.

4.4. Genes Encoding VF Detection by Polymerase Chain Reaction

All *E. coli* isolates were tested for the genes encoding VFs characteristic of pathogenic *E. coli* O157:H7: *stx1, stx2, rfbE, fliC* and *eaeA*, using specific primers [68]. Each *polymerase chain reaction* (PCR) reaction was performed in a 50 μL amplification mixture consisting of 10 μL 5 × PCR buffer (1.5 mM MgCl$_2$), 5.0 μL dNTPs (2.5 mM), 1 μL of each primer (10 μM), 0.25 μL of GoTaq DNA polymerase (5 unit/μL) and 10 μL of template. *E. coli* ATCC 43894 was used as a reference strain (*E. coli* O157:H7). The sequence of the used primers and the conditions of PCR were performed according to Tabashsum et al. [68]. Amplification products were separated by electrophoresis on 1.5% agarose gel, on 1 × Tris-Acetate-EDTA (242 g/L trizma base; 57.1 mL/L glacial acetic acid; 100 mL/L EDTA 0.5 M pH 8.0) at 100 V for 1 h and then visualized by GelRed staining, illuminated by UV transilluminator and visualized by a gel reader (Bio Rad Gel DOC XR+, Hercules, CA, USA). A 100 bp DNA ladder was used as a marker for PCR assay. The expected sizes of products for *eaeA, rfb* O157 and *fliC* H7 gene amplification were 150, 259 and 625 bp, and for *stx1* and *stx2* genes were 348 and 584 bp, respectively [68].

4.5. Antimicrobial Susceptibility Testing

Antimicrobial susceptibility of the isolates was tested using the Kirby Bauer method according to the Clinical and Laboratory Standards Institute (CLSI) guidelines [69]. The following antibiotics were tested: NA, 30 μg; CIP, 5 μg; AML, 25 μg; AUG, 20/10 μg; CTX, 30 μg; TE, 30 μg; SXT, 1,25/23,75 μg; N, 30 μg; C, 30 μg; CT, 50 μg (Oxoid, Hampshire, UK). Briefly, the isolates were grown on TSA for 24 h at 37 °C. Subsequently, each bacterial suspension was adjusted to McFarland 0.5 in normal saline and uniformly spread onto Mueller–Hinton agar (MHA; Oxoid, Hampshire, UK). Paper disks impregnated with antibiotics were placed on the surface of agar plates and incubated for 24 h at 37 °C aerobically. Then, the diameters of the inhibition zones were measured by using a Vernier caliper and the values were interpreted according to the CLSI guidelines [69]. *E. coli* ATCC 25922 and ATCC 10536 (American Type Culture Collection, Rockville, MD, USA) were used as quality control strains.

4.6. Phenotypic Confirmation of ESBL Production

Phenotypic confirmation of ESBL production was performed by double-disk synergy test according to the CLSI guidelines [69], by positioning an AUG disk at a distance of 30 mm to third-generation cephalosporin disk (CTX) on MHA. The test was considered as positive when a synergy (champagne cork aspect) between AUG and CTX disks was observed in combination with resistance or reduced susceptibility to third-generation cephalosporin. Isolates showing decreased susceptibility to third-generation cephalosporin without clear synergy were subjected to a Combination Disk Test, by applying disks containing third-generation cephalosporin alone and in combination with clavulanic acid, following CLSI guidelines [69].

4.7. ESBL Gene Identification by PCR

DNA of the isolated *E. coli* strains was prepared by boiling methods. Briefly, for each strain, 2 or 3 colonies were dissociated in 1 mL of distilled sterile water and centrifuged for 5 min at 13,000 rpm. The supernatant was eliminated, and the pellet was suspended in 200 µL of distilled sterile water and heated at 100 °C for 10 min, cooled on ice for 5 min, and the DNA was removed from the supernatant after 5 min of centrifugation (13,000 rpm) to pellet the cellular debris and stored at −20 °C until use. Genetic characterization of ESBLs was performed on phenotypically confirmed *E. coli* isolates by PCR. The sequence of primers and the conditions of PCR for the detection of bla_{ESBL} genes were performed as described previously for bla_{CTX-M} genotype groups 1, 2, 8 and 9, bla_{SHV} [70] and bla_{TEM} [71]. Amplification products were separated by gel electrophoresis using a 2% agarose gel.

4.8. Putative Enteroaggregative Gene Detection by PCR

The isolates were also investigated for the detection of various enteroaggregative putative genes: *aggr*, *astA* and *irp2*. The sequence of the used primers and the conditions of PCR were performed according to Mohamed et al. [9].

4.9. Biofilm Formation Assay

All *E. coli* isolates were evaluated for their ability to form biofilm by staining assay, as described by Cramton et al. [72] with some minor modifications. Briefly, overnight cultures in tryptic soy broth (TSB) were adjusted in culture medium to 5×10^5 CFU/mL and then 200 µL was dispensed into all the wells of the microtiter plate. The biofilm biomass formed in each well, after incubation for 24 h at 37 °C, was washed twice with phosphate-buffered saline (PBS), dried at room temperature, stained with aqueous 0.1% safranin solution (200 µL) for 1 min and then washed with water. The stained biofilms were dissolved in 30% (v/v) acetic acid and measured at OD 492 nm using a microplate reader. The following formula was applied to classify the biofilm formation: SBF = (AB − CW)/G, where AB is the stained attached bacteria (OD 492 nm), CW is the stained control wells containing bacteria-free medium only (OD 492 nm) and G is the cell growth in suspended culture (OD 540 nm) [73]. *E. coli* ATCC 25922 served as a positive control. TSB without bacteria was included as medium control.

The degree of biofilm formation of the isolates was classified into 4 categories: negative (SBF < 0.35), weak (SBF ≥ 0.35–0.69), moderate (SBF ≥ 0.70–1.09) and strong (SBF ≥ 1.10) [74].

4.10. Statistical Analysis

All experiments were performed in triplicate. Statistical data analysis was carried out using MATLAB_R2020a (MatWorks, Inc. Natick, MA, USA). A two-tailed Student's *t*-test was applied to evaluate the mean ± standard deviation and the significant differences in the grade of biofilm formation among different strains. For each comparison between virulence or resistance genes and biofilm formation, a correlation coefficient (r) was determined via Pearson's analysis. *p*-values of ≤0.05 were considered significant in all experiments.

5. Conclusions

In conclusion, our results reported a low frequency of virulence-associated genes of STEC O157:H7 in *E. coli* strains isolated from different poultry farms in Western Algeria. However, all isolates were shown to be resistant to at least three antibiotics most frequently used in poultry, and among these more than half were ESBL-positive *E. coli* despite no use of third-generation cephalosporins in Algerian poultry production. The ability of biofilm formation, which is considered a further virulent factor in pathogenic bacteria, was instead found to be higher among non-ESBL-producing strains of *E. coli*. Given that *E. coli* in chickens represents one of the major opportunistic pathogens and that it can be easily transferred from animals to humans, ESBL-producing *E. coli* represents an important risk factor for the poultry industry and human health. This study emphasizes the importance of monitoring the spread of the *E. coli* isolates that harbor virulence and antibiotic resistance genes in poultry farms, including the ones with healthy chickens, in order to prevent and control the spread of resistant bacteria and their virulence genes.

In Algeria, antimicrobials are not only used for therapeutic reasons but also for growth promotion and disease prevention. Consequently, the Algerian authorities should enforce AMR rules in order to guarantee a wise use of antimicrobials that will limit the risk of transmission along the food chain.

Author Contributions: Conceptualization, A.M., Q.B. and T.G.; Methodology, Q.B., T.G., F.G., D.A. and E.L.C.; Investigation, F.G., M.V., A.N., E.L.C. and N.C.; Writing—original draft preparation, A.M., Q.B. and T.G.; Data curation, A.N., N.C. and M.V.; Writing—review and editing, A.M. and T.G.; Funding acquisition, A.M. All authors have read and agreed to the published version of the manuscript.

Funding: This research received no external funding.

Institutional Review Board Statement: Not applicable.

Conflicts of Interest: The authors declare no conflict of interest.

References

1. Das, Q.; Islam, R.; Marcone, M.; Warriner, K.; Diarra, M. Potential of berry extracts to control foodborne pathogens. *Food Control* **2017**, *73*, 650–662. [CrossRef]
2. Gordon, D.M.; Cowling, A. The distribution and genetic structure of *Escherichia coli* in Australian vertebrates: Host and geographic effects. *Microbiology* **2003**, *149*, 3575–3586. [CrossRef]
3. Liu, C.M.; Stegger, M.; Aziz, M.; Johnson, T.J.; Waits, K.; Nordstrom, L.; Gauld, L.; Weaver, B.; Rolland, D.; Statham, S.; et al. *Escherichia coli* ST131-H22 as a Foodborne Uropathogen. *mBio* **2018**, *9*. [CrossRef] [PubMed]
4. Osman, K.M.; Kappell, A.D.; Elhadidy, M.; ElMougy, F.; El-Ghany, W.A.A.; Orabi, A.; Mubarak, A.S.; Dawoud, T.M.; Hemeg, H.A.; Moussa, I.M.I.; et al. Poultry hatcheries as potential reservoirs for antimicrobial-resistant *Escherichia coli*: A risk to public health and food safety. *Sci. Rep.* **2018**, *8*, 5859. [CrossRef] [PubMed]
5. Ramos, S.; Silva, V.; Dapkevicius, M.L.E.; Canica, M.; Tejedor-Junco, M.T.; Igrejas, G.; Poeta, P. *Escherichia coli* as Commensal and Pathogenic Bacteria Among Food-Producing Animals: Health Implications of Extended Spectrum beta-lactamase (ESBL) Production. *Animals* **2020**, *10*, 2239. [CrossRef] [PubMed]
6. Donnenberg, M.S. *Escherichia coli Virulence Mechanisms of a Versatile Pathogen*; Academic press: Baltimore, MD, USA; Elsevier Science: Amsterdam, The Netherlands, 2002.
7. Johnson, J.R.; Porter, S.B.; Zhanel, G.; Kuskowski, M.A.; Denamur, E. Virulence of *Escherichia coli* clinical isolates in a murine sepsis model in relation to sequence type ST131 status, fluoroquinolone resistance, and virulence genotype. *Infect. Immun.* **2012**, *80*, 1554–1562. [CrossRef]
8. Kaper, J.B.; Nataro, J.P.; Mobley, H.L. Pathogenic *Escherichia coli*. *Nat. Rev. Microbiol.* **2004**, *2*, 123–140. [CrossRef]
9. Mohamed, J.A.; Huang, D.B.; Jiang, Z.D.; DuPont, H.L.; Nataro, J.P.; Belkind-Gerson, J.; Okhuysen, P.C. Association of putative enteroaggregative *Escherichia coli* virulence genes and biofilm production in isolates from travelers to developing countries. *J. Clin. Microbiol.* **2007**, *45*, 121–126. [CrossRef]
10. Patel, J.; Yin, H.B.; Bauchan, G.; Mowery, J. Inhibition of *Escherichia coli* O157:H7 and *Salmonella enterica* virulence factors by benzyl isothiocyanate. *Food Microbiol.* **2020**, *86*, 103303. [CrossRef]
11. Hacker, J.; Kaper, J.B. Pathogenicity islands and the evolution of microbes. *Annu. Rev. Microbiol.* **2000**, *54*, 641–679. [CrossRef] [PubMed]
12. Huddleston, J.R. Horizontal gene transfer in the human gastrointestinal tract: Potential spread of antibiotic resistance genes. *Infect. Drug Resist.* **2014**, *7*, 167–176. [CrossRef]
13. Arabi, S.; Jafarpour, M.; Mirinargesi, M.; Asl, S.B.; Naghshbandi, R.; Shabanpour, M. Molecular characterization of avian pathogenic *Escherichia coli* in broilers bred in northern Iran. *Glob. Vet.* **2013**, *10*, 382–386.
14. Dissanayake, D.R.; Octavia, S.; Lan, R. Population structure and virulence content of avian pathogenic *Escherichia coli* isolated from outbreaks in Sri Lanka. *Vet. Microbiol.* **2014**, *168*, 403–412. [CrossRef] [PubMed]
15. Lindstedt, B.A.; Finton, M.D.; Porcellato, D.; Brandal, L.T. High frequency of hybrid Escherichia coli strains with combined Intestinal Pathogenic *Escherichia coli* (IPEC) and Extraintestinal Pathogenic *Escherichia coli* (ExPEC) virulence factors isolated from human faecal samples. *BMC Infect. Dis.* **2018**, *18*, 544. [CrossRef] [PubMed]
16. Maloo, A.; Fulke, A.B.; Khade, K.; Sharma, A.; Sukumaran, S. Virulence gene and antibiogram profile as markers of pathogenic *Escherichia coli* in tropical beaches of North Western India: Implications for water quality and human health. *Estuar. Coast. Shelf Sci.* **2018**, *208*, 118–130. [CrossRef]

17. Beninati, C.; Reich, F.; Muscolino, D.; Giarratana, F.; Panebianco, A.; Klein, G.; Atanassova, V. ESBL-Producing Bacteria and MRSA Isolated from Poultry and Turkey Products Imported from Italy. *Czech J. Food Sci.* **2015**, *33*, 2015–2097. [CrossRef]
18. Marotta, S.M.; Giarratana, F.; Calvagna, A.; Ziino, G.; Giuffrida, A.; Panebianco, A. Study on microbial communities in domestic kitchen sponges: Evidence of Cronobacter sakazakii and Extended Spectrum Beta Lactamase (ESBL) producing bacteria. *Ital. J. Food Saf.* **2019**, *7*, 7672. [CrossRef] [PubMed]
19. Marshall, B.M.; Levy, S.B. Food animals and antimicrobials: Impacts on human health. *Clin. Microbiol. Rev.* **2011**, *24*, 718–733. [CrossRef]
20. Overdevest, I.; Willemsen, I.; Rijnsburger, M.; Eustace, A.; Xu, L.; Hawkey, P.; Heck, M.; Savelkoul, P.; Vandenbroucke-Grauls, C.; van der Zwaluw, K.; et al. Extended-spectrum beta-lactamase genes of *Escherichia coli* in chicken meat and humans, The Netherlands. *Emerg. Infect. Dis.* **2011**, *17*, 1216–1222. [CrossRef] [PubMed]
21. van den Bogaard, A.E.; Stobberingh, E.E. Epidemiology of resistance to antibiotics. Links between animals and humans. *Int. J. Antimicrob. Agents* **2000**, *14*, 327–335. [CrossRef]
22. Arias, C.A.; Murray, B.E. Antibiotic-resistant bugs in the 21st century—A clinical super-challenge. *N. Engl. J. Med.* **2009**, *360*, 439–443. [CrossRef]
23. Pitout, J.D.D.; Laupland, K.B. Extended-spectrum β-lactamase-producing Enterobacteriaceae: An emerging public-health concern. *Lancet Infect. Dis.* **2008**, *8*, 159–166. [CrossRef]
24. Chong, Y.; Ito, Y.; Kamimura, T. Genetic evolution and clinical impact in extended-spectrum beta-lactamase-producing *Escherichia coli* and *Klebsiella pneumoniae*. *Infect. Genet. Evol.* **2011**, *11*, 1499–1504. [CrossRef]
25. Sanjit Singh, A.; Lekshmi, M.; Prakasan, S.; Nayak, B.B.; Kumar, S. Multiple Antibiotic-Resistant, Extended Spectrum-beta-Lactamase (ESBL)-Producing Enterobacteria in Fresh Seafood. *Microorganisms* **2017**, *5*, 53. [CrossRef] [PubMed]
26. Paterson, D.L.; Bonomo, R.A. Extended-spectrum beta-lactamases: A clinical update. *Clin. Microbiol. Rev.* **2005**, *18*, 657–686. [CrossRef] [PubMed]
27. Abdallah, M.; Benoliel, C.; Drider, D.; Dhulster, P.; Chihib, N.E. Biofilm formation and persistence on abiotic surfaces in the context of food and medical environments. *Arch. Microbiol.* **2014**, *196*, 453–472. [CrossRef] [PubMed]
28. Colagiorgi, A.; Bruini, I.; Di Ciccio, P.A.; Zanardi, E.; Ghidini, S.; Ianieri, A. *Listeria monocytogenes* Biofilms in the Wonderland of Food Industry. *Pathogens* **2017**, *6*, 41. [CrossRef]
29. Galie, S.; Garcia-Gutierrez, C.; Miguelez, E.M.; Villar, C.J.; Lombo, F. Biofilms in the Food Industry: Health Aspects and Control Methods. *Front. Microbiol.* **2018**, *9*, 898. [CrossRef]
30. Mellata, M.; Johnson, J.R.; Curtiss, R., 3rd. *Escherichia coli* isolates from commercial chicken meat and eggs cause sepsis, meningitis and urinary tract infection in rodent models of human infections. *Zoonoses Public Health* **2018**, *65*, 103–113. [CrossRef]
31. Stromberg, Z.R.; Johnson, J.R.; Fairbrother, J.M.; Kilbourne, J.; Van Goor, A.; Curtiss, R.R.; Mellata, M. Evaluation of *Escherichia coli* isolates from healthy chickens to determine their potential risk to poultry and human health. *PLoS ONE* **2017**, *12*, e0180599. [CrossRef]
32. Rouger, A.; Tresse, O.; Zagorec, M. Bacterial Contaminants of Poultry Meat: Sources, Species, and Dynamics. *Microorganisms* **2017**, *5*, 50. [CrossRef]
33. Alloui, N.; Bennoune, O. Poultry production in Algeria: Current situation and future prospects. *Worlds Poult. Sci. J.* **2013**, *69*, 613–620. [CrossRef]
34. Baliere, C.; Rince, A.; Delannoy, S.; Fach, P.; Gourmelon, M. Molecular Profiling of Shiga Toxin-Producing *Escherichia coli* and Enteropathogenic *E. coli* Strains Isolated from French Coastal Environments. *Appl. Environ. Microbiol.* **2016**, *82*, 3913–3927. [CrossRef] [PubMed]
35. Wang, J.; Zhao, G.; Gao, Y.; Xu, H.; Mohamed, L.; Zhao, J.; Gai, W.; Zou, M.; Cui, Z.; Yan, S.; et al. Virulence and Antimicrobial Characteristics of *Escherichia coli* Isolated from Diseased Chickens in China and Algeria. *J. Adv. Agric.* **2019**, *10*, 1821–1833. [CrossRef]
36. Mellata, M.; Bakour, R.; Jacquemin, E.; Mainil, J.G. Genotypic and phenotypic characterization of potential virulence of intestinal avian *Escherichia coli* strains isolated in Algeria. *Avian Dis.* **2001**, *45*, 670–679. [CrossRef] [PubMed]
37. Doane, C.A.; Pangloli, P.; Richards, H.A.; Mount, J.R.; Golden, D.A.; Draughon, F.A. Occurrence of *Escherichia coli* O157:H7 in diverse farm environments. *J. Food Prot.* **2007**, *70*, 6–10. [CrossRef]
38. Ferens, W.A.; Hovde, C.J. *Escherichia coli* O157:H7: Animal reservoir and sources of human infection. *Foodborne Pathog. Dis.* **2011**, *8*, 465–487. [CrossRef]
39. Koochakzadeh, A.; Badouei, M.A.; Salehi, T.Z.; Aghasharif, S.; Soltani, M.; Ehsan, M.R. Prevalence of Shiga Toxin-Producing and Enteropathogenic *Escherichia coli* in Wild and Pet Birds in Iran. *Braz. J. Poult. Sci.* **2015**, *14*, 5. [CrossRef]
40. Persad, A.K.; LeJeune, J.T. Animal Reservoirs of Shiga Toxin-Producing *Escherichia coli*. *Microbiol. Spectr.* **2014**, *2*, EHEC-0027-2014. [CrossRef]
41. Poole, T.; Sheffield, C. Use and misuse of antimicrobial drugs in poultry and livestock: Mechanisms of antimicrobial resistance. *Pak. Vet. J.* **2013**, *33*, 266–271.
42. Johnson, T.J.; Logue, C.M.; Johnson, J.R.; Kuskowski, M.A.; Sherwood, J.S.; Barnes, H.J.; DebRoy, C.; Wannemuehler, Y.M.; Obata-Yasuoka, M.; Spanjaard, L.; et al. Associations between multidrug resistance, plasmid content, and virulence potential among extraintestinal pathogenic and commensal *Escherichia coli* from humans and poultry. *Foodborne Pathog. Dis.* **2012**, *9*, 37–46. [CrossRef]

43. Mellata, M. Human and avian extraintestinal pathogenic *Escherichia coli*: Infections, zoonotic risks, and antibiotic resistance trends. *Foodborne Pathog. Dis.* **2013**, *10*, 916–932. [CrossRef]

44. Castanon, J.I. History of the use of antibiotic as growth promoters in European poultry feeds. *Poult. Sci.* **2007**, *86*, 2466–2471. [CrossRef]

45. Agersø, Y.; Jensen, J.D.; Hasman, H.; Pedersen, K. Spread of extended spectrum cephalosporinase-producing *Escherichia coli* clones and plasmids from parent animals to broilers and to broiler meat in a production without use of cephalosporins. *Foodborne Pathog. Dis.* **2014**, *11*, 740–746. [CrossRef]

46. Bondt, N.; Puister-Jansen, L.F. *MARAN 2004: Monitoring of Antimicrobial Resistance and Antibiotic Usage in Animals in the Netherlands in 2004*; CIDC-Lelystad, LEI Wageningen UR, VWA, RIVM: Lelystad, The Netherlands, 2005.

47. Boutaiba Benklaouz, M.; Aggad, H.; Benameur, Q. Resistance to multiple first-line antibiotics among *Escherichia coli* from poultry in Western Algeria. *Vet. World* **2020**, *13*, 290–295. [CrossRef]

48. Hussain, A.; Ranjan, A.; Nandanwar, N.; Babbar, A.; Jadhav, S.; Ahmed, N. Genotypic and phenotypic profiles of *Escherichia coli* isolates belonging to clinical sequence type 131 (ST131), clinical non-ST131, and fecal non-ST131 lineages from India. *Antimicrob. Agents Chemother.* **2014**, *58*, 7240–7249. [CrossRef]

49. Nandanwar, N.; Hussain, A.; Ranjan, A.; Jadhav, S.; Ahmed, N. Population structure and molecular epidemiology of human clinical multi-drug resistant (MDR) *Escherichia coli* strains from Pune, India. *Int. J. Infect. Dis.* **2016**, *45*, 343–344. [CrossRef]

50. Shaik, S.; Kumar, N.; Lankapalli, A.K.; Tiwari, S.K.; Baddam, R.; Ahmed, N. Contig-Layout-Authenticator (CLA): A Combinatorial Approach to Ordering and Scaffolding of Bacterial Contigs for Comparative Genomics and Molecular Epidemiology. *PLoS ONE* **2016**, *11*, e0155459. [CrossRef] [PubMed]

51. Dierikx, C.M.; van der Goot, J.A.; Smith, H.E.; Kant, A.; Mevius, D.J. Presence of ESBL/AmpC-producing *Escherichia coli* in the broiler production pyramid: A descriptive study. *PLoS ONE* **2013**, *8*, e79005. [CrossRef]

52. Mo, S.S.; Slettemeås, J.S.; Berg, E.S.; Norström, M.; Sunde, M. Plasmid and Host Strain Characteristics of *Escherichia coli* Resistant to Extended-Spectrum Cephalosporins in the Norwegian Broiler Production. *PLoS ONE* **2016**, *11*, e0154019. [CrossRef] [PubMed]

53. Belmahdi, M.; Bakour, S.; Al Bayssari, C.; Touati, A.; Rolain, J.M. Molecular characterisation of extended-spectrum beta-lactamase- and plasmid AmpC-producing *Escherichia coli* strains isolated from broilers in Bejaia, Algeria. *J. Glob. Antimicrob. Resist.* **2016**, *6*, 108–112. [CrossRef] [PubMed]

54. Benameur, Q.; Tali-Maamar, H.; Assaous, F.; Guettou, B.; Rahal, K.; Ben-Mahdi, M.H. Detection of multidrug resistant *Escherichia coli* in the ovaries of healthy broiler breeders with emphasis on extended-spectrum beta-lactamases producers. *Comp. Immunol. Microbiol. Infect. Dis.* **2019**, *64*, 163–167. [CrossRef]

55. Meguenni, N.; Le Devendec, L.; Jouy, E.; Le Corvec, M.; Bounar-Kechih, S.; Rabah Bakour, D.; Kempf, I. First Description of an Extended-Spectrum Cephalosporin- and Fluoroquinolone- Resistant Avian Pathogenic *Escherichia coli* Clone in Algeria. *Avian Dis.* **2015**, *59*, 20–23. [CrossRef]

56. Tekiner, I.H.; Ozpinar, H. Occurrence and characteristics of extended spectrum beta-lactamases-producing Enterobacteriaceae from foods of animal origin. *Braz. J. Microbiol.* **2016**, *47*, 444–451. [CrossRef]

57. Sah, R.S.P.; Dhungel, B.; Yadav, B.K.; Adhikari, N.; Thapa Shrestha, U.; Lekhak, B.; Banjara, M.R.; Adhikari, B.; Ghimire, P.; Rijal, K.R. Detection of TEM and CTX-M Genes in *Escherichia coli* Isolated from Clinical Specimens at Tertiary Care Heart Hospital, Kathmandu, Nepal. *Diseases* **2021**, *9*, 15. [CrossRef]

58. Lee, S.-Y.; Park, Y.-J.; Yu, J.K.; Jung, S.; Kim, Y.; Jeong, S.H.; Arakawa, Y. Prevalence of acquired fosfomycin resistance among extended-spectrum β-lactamase-producing *Escherichia coli* and *Klebsiella pneumoniae* clinical isolates in Korea and IS26-composite transposon surrounding fosA3. *J. Antimicrob. Chemother.* **2012**, *67*, 2843–2847. [CrossRef] [PubMed]

59. Bradford, P.A. Extended-spectrum beta-lactamases in the 21st century: Characterization, epidemiology, and detection of this important resistance threat. *Clin. Microbiol. Rev.* **2001**, *14*, 933–951. [CrossRef]

60. Habeeb, M.A.; Haque, A.; Iversen, A.; Giske, C.G. Occurrence of virulence genes, 16S rRNA methylases, and plasmid-mediated quinolone resistance genes in CTX M producing *Escherichia coli* from Pakistan. *Eur. J. Clin. Microbiol. Infect. Dis.* **2014**, *33*, 399–409. [CrossRef]

61. Projahn, M.; Daehre, K.; Semmler, T.; Guenther, S.; Roesler, U.; Friese, A. Environmental adaptation and vertical dissemination of ESBL-/pAmpC-producing *Escherichia coli* in an integrated broiler production chain in the absence of an antibiotic treatment. *Microb. Biotechnol.* **2018**, *11*, 1017–1026. [CrossRef] [PubMed]

62. Wang, R. Biofilms and Meat Safety: A Mini-Review. *J. Food Prot.* **2019**, *82*, 120–127. [CrossRef] [PubMed]

63. Flemming, H.C.; Wingender, J. The biofilm matrix. *Nat. Reviews. Microbiol.* **2010**, *8*, 623–633. [CrossRef]

64. Hall-Stoodley, L.; Costerton, J.W.; Stoodley, P. Bacterial biofilms: From the natural environment to infectious diseases. *Nat. Reviews. Microbiol.* **2004**, *2*, 95–108. [CrossRef]

65. Gallant, C.V.; Daniels, C.; Leung, J.M.; Ghosh, A.S.; Young, K.D.; Kotra, L.P.; Burrows, L.L. Common beta-lactamases inhibit bacterial biofilm formation. *Mol. Microbiol.* **2005**, *58*, 1012–1024. [CrossRef]

66. Sarantuya, J.; Nishi, J.; Wakimoto, N.; Erdene, S.; Nataro, J.P.; Sheikh, J.; Iwashita, M.; Manago, K.; Tokuda, K.; Yoshinaga, M.; et al. Typical enteroaggregative *Escherichia coli* is the most prevalent pathotype among *E. coli* strains causing diarrhea in Mongolian children. *J. Clin. Microbiol.* **2004**, *42*, 133–139. [CrossRef]

67. Sheikh, J.; Hicks, S.; Dall'Agnol, M.; Phillips, A.D.; Nataro, J.P. Roles for Fis and YafK in biofilm formation by enteroaggregative *Escherichia coli*. *Mol. Microbiol.* **2001**, *41*, 983–997. [CrossRef] [PubMed]

68. Tabashsum, Z.; Nazneen, M.; Ahsan, C.R.; Bari, M.L.; Yasmin, M. Influence of Detection Methods in Characterizing *Escherichia coli* O157:H7 in Raw Goat Meat Using Conventional and Molecular Methods. *Biocontrol Sci.* **2016**, *21*, 261–264. [CrossRef]
69. Clinical and Laboratory Standards Institute (CLSI). *Performance Standards for Antimicrobial Disk Susceptibility Tests*, 27th ed.; Approved Standard; CLSI standard: Wayne, PA, USA, 2017.
70. Arlet, G.; Rouveau, M.; Philippon, A. Substitution of alanine for aspartate at position 179 in the SHV-6 extended-spectrum beta-lactamase. *FEMS Microbiol. Lett.* **1997**, *152*, 163–167. [CrossRef]
71. Mabilat, C.; Goussard, S. PCR detection and identification of genes for extended-spectrum beta-lactamases. In *Diagnostic Molecular Microbiology: Principles and Applications. American Society for Microbiology*; Persing, D.H., Smith, T.F., Tenover, T.C., White, T.J., Eds.; American Society for Microbiology: Washington, DC, USA, 1993; pp. 553–559.
72. Cramton, S.E.; Gerke, C.; Schnell, N.F.; Nichols, W.W.; Götz, F. The intercellular adhesion (ica) locus is present in *Staphylococcus aureus* and is required for biofilm formation. *Infect. Immun.* **1999**, *67*, 5427–5433. [CrossRef] [PubMed]
73. Niu, C.; Gilbert, E.S. Colorimetric method for identifying plant essential oil components that affect biofilm formation and structure. *Appl. Environ. Microbiol.* **2004**, *70*, 6951–6956. [CrossRef]
74. Naves, P.; del Prado, G.; Huelves, L.; Gracia, M.; Ruiz, V.; Blanco, J.; Rodriguez-Cerrato, V.; Ponte, M.C.; Soriano, F. Measurement of biofilm formation by clinical isolates of *Escherichia coli* is method-dependent. *J. Appl. Microbiol.* **2008**, *105*, 585–590. [CrossRef] [PubMed]

antibiotics

Article

Climatic Alterations Influence Bacterial Growth, Biofilm Production and Antimicrobial Resistance Profiles in *Aeromonas* spp.

Miguel L. Grilo [1,2,*], **Ana Pereira** [2], **Carla Sousa-Santos** [2], **Joana I. Robalo** [2] and **Manuela Oliveira** [1,*]

[1] Centro de Investigação Interdisciplinar em Sanidade Animal (CIISA), Faculdade de Medicina Veterinária, Universidade de Lisboa, 1300-477 Lisbon, Portugal

[2] Marine and Environmental Sciences Centre (MARE), Instituto Universitário de Ciências Psicológicas, Sociais e da Vida (ISPA), 1100-304 Lisbon, Portugal; ana_pereira@ispa.pt (A.P.); carla.santos@ispa.pt (C.S.-S.); jrobalo@ispa.pt (J.I.R.)

* Correspondence: miguelgrilo@fmv.ulisboa.pt (M.L.G.); moliveira@fmv.ulisboa.pt (M.O.)

Abstract: Climate change is expected to create environmental disruptions that will impact a wide array of biota. Projections for freshwater ecosystems include severe alterations with gradients across geographical areas. Life traits in bacteria are modulated by environmental parameters, but there is still uncertainty regarding bacterial responses to changes caused by climatic alterations. In this study, we used a river water microcosm model to evaluate how *Aeromonas* spp., an important pathogenic and zoonotic genus ubiquitary in aquatic ecosystems, responds to environmental variations of temperature and pH as expected by future projections. Namely, we evaluated bacterial growth, biofilm production and antimicrobial resistance profiles of *Aeromonas* species in pure and mixed cultures. Biofilm production was significantly influenced by temperature and culture, while temperature and pH affected bacterial growth. Reversion of antimicrobial susceptibility status occurred in the majority of strains and tested antimicrobial compounds, with several combinations of temperature and pH contributing to this effect. Current results highlight the consequences that bacterial genus such as *Aeromonas* will experience with climatic alterations, specifically how their proliferation and virulence and phenotypic resistance expression will be modulated. Such information is fundamental to predict and prevent future outbreaks and deleterious effects that these bacterial species might have in human and animal populations.

Keywords: microcosm; *Aeromonas*; climate change; temperature; pH; biofilm; antimicrobial resistance; water

Citation: Grilo, M.L.; Pereira, A.; Sousa-Santos, C.; Robalo, J.I.; Oliveira, M. Climatic Alterations Influence Bacterial Growth, Biofilm Production and Antimicrobial Resistance Profiles in *Aeromonas* spp. *Antibiotics* **2021**, *10*, 1008. https://doi.org/10.3390/antibiotics10081008

Academic Editors: Carlos M. Franco and Marc Maresca

Received: 14 July 2021
Accepted: 17 August 2021
Published: 20 August 2021

Publisher's Note: MDPI stays neutral with regard to jurisdictional claims in published maps and institutional affiliations.

1. Introduction

Environmental conditions are a major driver of bacterial activity and can shape the expression of several metabolic pathways [1,2]. Namely, such parameters have the potential to influence bacterial virulence (e.g., biofilm formation) and antibiotic resistance signatures [3,4].

Climatic scenarios, as predicted by simulation methodologies based on different levels of emissions, are projected to significantly differ from currently observed meteorological conditions [5]. Regarding aquatic ecosystems, and particularly in freshwater habitats, various environmental parameters are expected to be altered in the coming years. Water temperature, directly influenced by air temperature, is expected to rise across different habitats [6]. Additionally, the occurrence of heatwaves will likely increase, resulting in extended periods of drought associated with a low flow of freshwater systems, a decrease in water level and in dissolved oxygen concentrations [7–9]. Consequently, reduced dilution of freshwater streams will also affect ion balance levels [10,11]. These biotic changes will impact ecosystem dynamics and promote disruptions in species equilibrium [9,12]. All of these events are expected to significantly decrease freshwater's quality [13]. Ultimately, these changes compromise future water availability, freshwater ecosystems' structure and populations' sustainability [14–16].

Natural aquatic ecosystems, often the last destination of terrestrial runoffs, are known reservoirs of both antimicrobial resistance and bacterial virulence determinants [17]. The microbiota present there, with or without direct connection with clinical infections, constitute a pool of information to the terrestrial microbiota or can even be disseminated to anthropogenic cycles [18]. This intricate connection between environmental microbiota and bacterial genus with effects at the One Health level stresses the importance of close surveillance of antimicrobial resistance and virulence dynamics in natural habitats in order to prevent epidemic situations both in anthropogenic settings and natural habitats [19,20]. Since modeling bacterial responses to changing environmental parameters in natural habitats is challenging, lab simulations—e.g., microcosm assays—are an important tool to predict how microbiota will respond to environmental cues foreseen in climatic predictions [21,22].

We hypothesize that aquatic bacteria's antimicrobial resistance signatures and virulence traits, as well as their growth, may vary with changing environmental conditions. In order to test this, we applied microcosm simulation assays using different water temperatures and pH values following established emissions scenarios [5] to *Aeromonas* spp.—a model bacterial genus ubiquitous across different aquatic ecosystems—and evaluated changes in the antimicrobial resistance profile, biofilm production and growth of the isolates under study.

2. Results

Biofilm production by each of the *Aeromonas* strains in pure and mixed culture in the different assays is illustrated in Figure 1. Each strain's response to temperature and pH was variable between species and within the same species.

When considering results by groups (*Aeromonas* species individually and mixed cultures), biofilm production in the mixed cultures' wells was significantly lower ($p < 0.001$) than in the other groups. Additionally, water temperature also significantly influenced biofilm production ($p = 0.006$), with isolates exposed to the Fluctuations treatment producing less biofilm (Figure 2). The different pH conditions tested did not influence biofilm production.

Regarding mixed culture wells, re-isolation and identification of the initial *Aeromonas* pool added to each well was not possible with several combinations of temperature and pH treatments. *Aeromonas* species prevalence at the end of microcosm assays varied across the applied treatments and also between replicates (Figure 3). When evaluating the influence of each individual *Aeromonas* species present in mixed cultures on the biofilm production, it was observed that no species had a significantly different influence.

Some differences were observed regarding the growth of the isolates during the experiment (Figure 1). Significant differences were recorded between the tested *Aeromonas* species ($p < 0.001$). *A. veronii* isolates presented significantly lower concentrations than the other single and mixed cultures, while *A. hydrophila* presented significantly lower concentrations than *A. media* and mixed cultures. Temperature ($p < 0.001$) and pH ($p = 0.007$) treatments also influenced bacterial growth. While bacterial growth did not differ between current and fluctuations treatments, it was significantly increased in the RCP 4.5 treatment and decreased in the RCP 8.5 treatment. Bacterial growth was increased in acidic pH conditions (6.31) when compared to alkaline pH (8.61). Specific associations were also found between *Aeromonas* species and pH ($p = 0.002$) and between temperature and pH ($p < 0.001$). While *A. media* and mixed cultures presented higher concentrations in water microcosms with pH 6.31, *A. caviae* presented higher concentrations at pH 8.61. No differences were observed at pH 7.61. Regarding the interaction between temperature and pH, concentrations in the RCP 4.5 treatment were higher at pH 6.31, decreasing until pH 8.61. For RCP 8.5, higher concentrations were observed at pH 8.61.

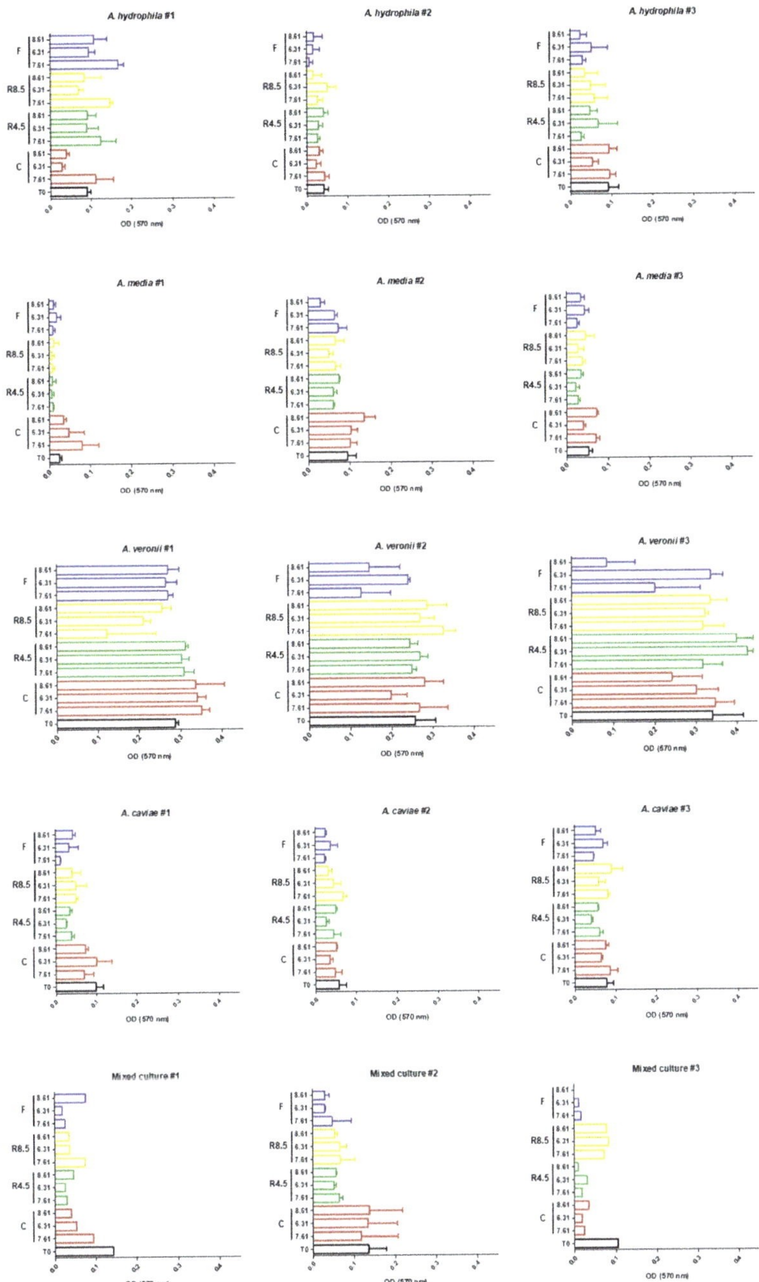

Figure 1. Biofilm production by each strain and mixed culture (mean + SEM). Presented results correspond to values subtracted to each treatment's negative control for normalization. The three replicates' results are presented by strain and mixed cultures, except for replicates where T0 and Current pH 7.61 were considered significantly different (*A. veronii* #1, Mixed cultures #1, #2 and #3). First column in each graph represents the temperature treatment (C—Current, R4.5—RCP 4.5, R8.5—RCP 8.5, F—Fluctuations) and the second the pH treatment. OD—Optical density.

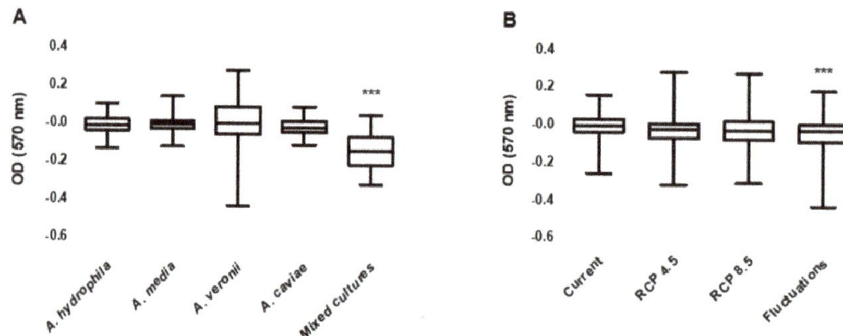

Figure 2. Biofilm production by individual *Aeromonas* species and mixed cultures (**A**) and in different water temperature treatment—each treatment includes all strains results. (**B**). Presented results correspond to values subtracted to each treatment's negative control for normalization and to the corresponding T0 treatment values for comparison. OD—Optical density; *** $p < 0.001$.

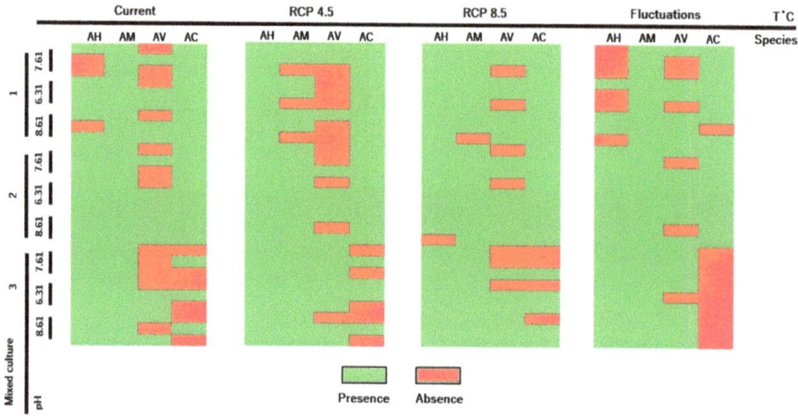

Figure 3. Prevalence of *Aeromonas* species in the mixed cultured wells after the microcosm assay. Each line corresponds to a distinct replicate belonging to one of the pH treatments (7.61, 6.31 and 8.61) from the tested mixed cultures (#1, #2 and #3). Each column represents an *Aeromonas* species (AH—*A. hydrophila*, AM—*A. media*, AV—*A. veronii*, AC—*A. caviae*) from a specific temperature treatment (Current, RCP 4.5, RCP 8.5 and Fluctuations).

The bacterial concentration was not correlated with biofilm production ($r_s = 0.020$, $p = 0.676$).

Several changes regarding the antimicrobial resistance profile were observed among treatments for the same isolate (Figure 5). Observations between the control treatment (T0, pH 7.61) were similar to results obtained with the current treatment and similar pH levels. Phenotype variation occurred in a strain-dependent way, and it was specific for each antimicrobial compound tested. For all strains and antibiotics (except *A. hydrophila* and tetracycline), modification of the original susceptibility category occurred with at least one combination of treatments.

In certain situations, reversion of non-wild-type to a wild-type phenotype occurred only with specific combinations of temperature and pH. This is the case of erythromycin susceptibility and *A. caviae*, *A. hydrophila* and *A. media*. Regarding *A. caviae* and *A. media*, the same treatment (i.e., Current and pH 6.31) caused this phenomenon. In other cases, several combinations resulted in this reversion with no obvious pattern. The opposite was also observed (conversion from wild-type to non-wild-type) among the isolates. Although some treatments seemed to result in this situation more often for some antimicrobial compounds (i.e., RCP 4.5), a high variability was observed.

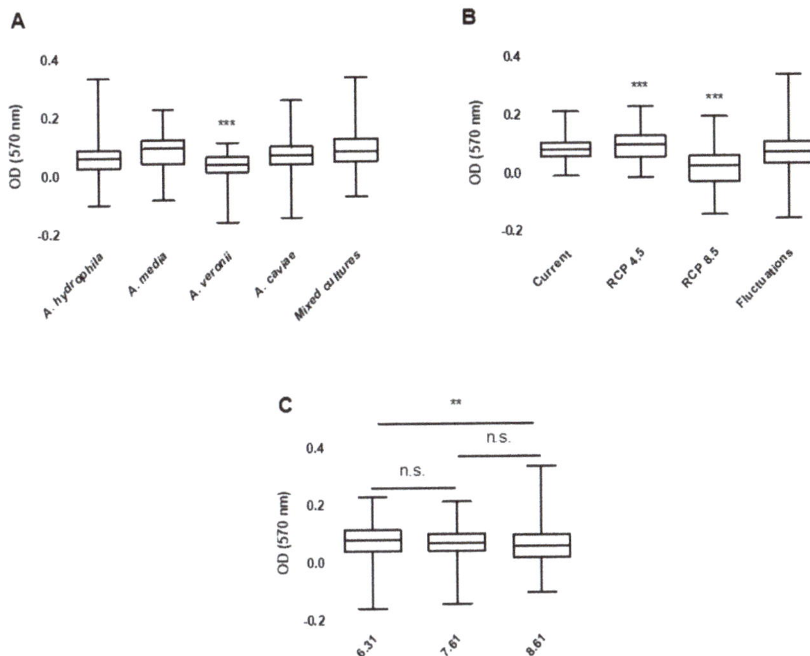

Figure 4. Bacterial concentration by *Aeromonas* species and mixed cultures (**A**), by water temperature treatment (**B**) and by water pH treatments (**C**). Presented results correspond to values subtracted to each treatment's negative control for normalization and to the corresponding T0 treatment values for comparison. OD—Optical density. n.s. $p > 0.05$, ** $p < 0.01$, *** $p < 0.001$.

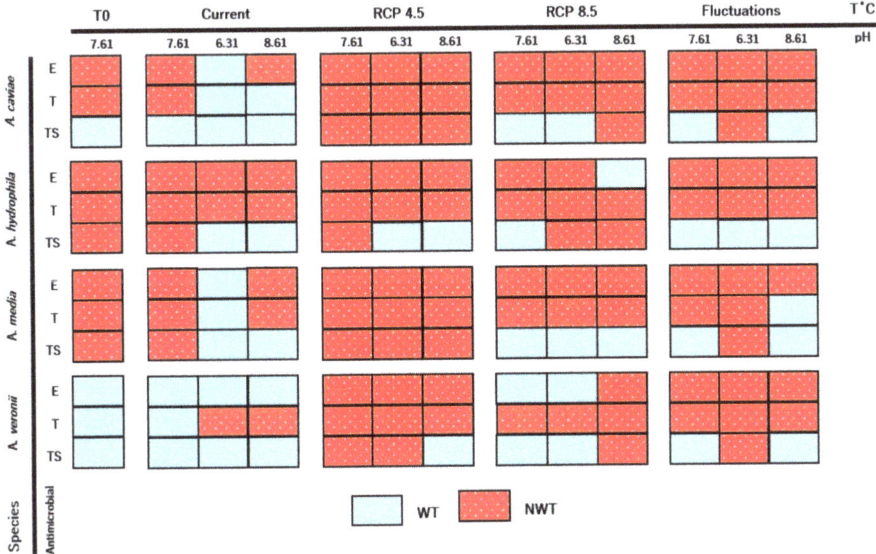

Figure 5. Antimicrobial resistance phenotypes (WT—wild-type, susceptible; NWT—non-wild-type, non-susceptible) of the *Aeromonas* isolates regarding water temperature and pH treatments. E—erythromycin, T—tetracycline, TS—sulfamethoxazole/trimethoprim.

3. Discussion

Investigating how bacteria will evolve with environmental cues using natural habitats is a difficult task. Instead, the use of microcosm simulations allows the exploration of such associations, ensuring experimental control and uniformity. This methodology represents a first step in the prediction of transformations to occur in important bacterial genus with an impact at the One Health level, such as *Aeromonas* spp., and prepare for future outbreaks or phenotypical changes with consequences to public health. In this study, we show that different *Aeromonas* species adapt their growth, biofilm production and antimicrobial resistance signatures to environmental projections related to climatic alterations (i.e., temperature and pH) in water, highlighting the role that future climatic events will have in shaping bacterial activity, as well as virulence and resistance expression. It is noteworthy that, in this study, differences regarding growth, biofilm production and antimicrobial resistance signatures were observed using relatively small temperature and pH amplitudes, which are more likely to reflect future climatic trends.

3.1. Biofilm Production

In general, the studied isolates presented variability in the production of biofilm when exposed to the different temperature and pH treatments. Although some response patterns were present, the disparity in results between isolates of different species and within the same species highlights the fact that individual characteristics will govern how an isolate will respond to environmental cues; however, significant associations were observed. Mixed cultures produced significantly less biofilm when compared to the *Aeromonas* species individually. At the end of the microcosm assay, it was not possible to isolate all *Aeromonas* species in many mixed cultures. Some species absence was more evident than others (e.g., *A. caviae* in mixed culture #3 along the various temperature and pH treatments), although a general pattern was not present. Additionally, and while pH treatments seem not to influence biofilm production significantly, temperature influenced biofilm production in *Aeromonas* spp. Namely, isolates exposed to temperature oscillations (i.e., Fluctuations) produced less biofilm. Such biofilm production was not dependent on bacterial concentration. Distinct *Aeromonas* species display specific preferences regarding environmental parameters [23,24]. Although *Aeromonas* spp. possess stress response mechanisms to deal with environmental oscillations [25], they still impact several aspects of bacterial life. If the combined temperature and pH conditions fall within the optimal range for multiplication and virulence expression for each isolate, they will dictate the isolate's competitiveness and ability to survive in an environment composed of multiple species [26,27]. Further, the level of interspecific competition for the limited resources will also hinder each isolate's ability to allocate nutrients to processes such as biofilm production, contrary to what occurs in pure cultures [28]. Finally, environmental oscillations of abiotic factors, such as temperature, will create additional disturbances for the bacterial communities [22] and the overall combination of external stressors with internal competition is likely to impact the final biofilm production.

3.2. Bacterial Growth

In this study, a disparity in bacterial growth during the microcosm experiments was observed between the studied species (both in pure culture and in mixed culture). Overall, *A. veronii* isolates displayed a significantly lower growth when compared with other tested groups. Delamare et al. [29] highlighted lower growth patterns by *A. veronii* when compared to other *Aeromonas* species (i.e., *A. hydrophila*, *A. media* and *A. caviae*). Growth rate variability is a consequence of phenotypic diversity in bacteria [30]. Such variability can be the result of the nutrient uptake rate by the bacterial cell and of the resource's distribution between the processes occurring in the bacterial cell [31]. *Aeromonas* strains and species growth variability likely reflect different limitations in these processes among the isolates, which can also explain differences observed not only for *A. veronii* but also for *A. hydrophila*. In the mixed cultures group, this pattern was not observed, and two hypotheses can be drawn: either other *Aeromonas* species present in the culture compensated for lower growth rates by *A. veronii*, or interspecific competition eliminated *A. veronii* presence in the microcosm wells (as stated in Figure 3), facilitating the growth of other species or canceling the growth effect *A. veronii* had in the total growth.

Both changes in water temperature, as well as in the pH conditions, played a significant role in the growth of *Aeromonas* spp. Regarding temperature, a biphasic effect was observed: while small increments in water temperature (i.e., RCP 4.5) seem to benefit the *Aeromonas* species under study, both in pure and in mixed cultures and favor their proliferation; once reaching a certain threshold imposed by higher temperatures (i.e., RCP 8.5), such boosting effect is lost and bacterial growth is lowered. Temperature is a determinant in bacterial growth and *Aeromonas* typically increase both growth and metabolic activity and decrease lag phase when experiencing higher environmental temperatures [32,33]; however, such growth reaches a plateau with temperature increments and starts to decrease before reaching maximum thermal tolerance [34], highlighting the role of thermal stress as a regulator of bacterial growth. It is noteworthy that, although cultures subjected to the fluctuation treatment experienced similar temperature values, such as the ones in the RCP 8.5 treatment, alternate exposure to higher (24.5 °C) and lower (21 °C) temperature values likely created buffer periods in which bacterial cultures could stabilize and multiplicate.

Regarding pH, the overall growth of *Aeromonas* spp. was higher in acidic environments when compared to alkaline environments. While some authors found a non-significant effect or a negative effect of pH on *Aeromonas* growth [33,35,36], *Aeromonas* are evolutionarily adapted to low pH environments, such as the gastrointestinal environment, and have built cellular responses (i.e., protective protein synthesis) that allow for acid tolerance [37]. Additionally, when exposed to acidic environments, the lag phase in *Aeromonas* is significantly shorter, prompting the beginning of the following growth phases sooner [38]; however, it is likely that different *Aeromonas* species display specific niche preferences and have evolved towards tolerance in different pH gradients. This explains why in this study some groups exhibited higher growths in acidic treatments (*A. media* and mixed cultures), while others performed better in alkaline pH (*A. caviae*). Additionally, both temperature and pH seem to play an interactive role, conditioning higher growth of *Aeromonas* spp. with specific combinations (i.e., RCP 4.5 and acidic pH, RCP 8.5 and alkaline pH).

3.3. Antimicrobial Resistance Profiles

Climate change has been implicated as a factor involved in increasing levels of antimicrobial resistance among different bacterial species in prolonged temporal sets. Distinct spatial patterns occur globally and are connected with local climacteric variability, highlighting how distinct geographical areas will be impacted by this problem in different proportions [39,40]. Specifically, regions expected to be more vulnerable to climacteric alterations are also the ones predicted to accumulate the highest prevalence of antimicrobial resistance [41]. Some authors report the role of increasing temperatures over time in the overexpression of this phenomenon in species such as *Escherichia coli*, *Klebsiella pneumoniae*, *Pseudomonas aeruginosa* and *Staphylococcus aureus* [40,42]. In a meta-analysis with isolates collected in aquacultures conducted by Reverter et al. [41], a similar conclusion was drawn for bacterial genera commonly infecting aquatic animals. In this study, we show that climatic scenarios of changing temperature and pH can alter the antimicrobial susceptibility profile of different *Aeromonas* species. Although species belonging to the *Aeromonas* genus are normally resistant to erythromycin and susceptible to tetracycline and sulfamethoxazole/trimethoprim, the selected strains in this study displayed variable susceptibility status to these antimicrobials; however, and with the exception of one strain (*A. hydrophila* and tetracycline), reversion of the original susceptibility status occurred for all tested strains and antimicrobial compounds at least in one experimental condition.

In some situations, reversion of non-susceptibility to susceptibility to the tested antimicrobial compounds was observed. Antibiotic resistance represents a fitness cost for bacterial species and the development of resistance is modulated by this parameter [43,44].

Resistance to antimicrobial compounds can impact important cellular activities or be met with higher energetic costs related to gene expression needs [45,46]. Thus, when experiencing amplified fitness costs, such as those provided by changes in temperature and pH, the rate of resistance reversibility in bacteria increases [43]. In this study, it seems that several combinations of water temperature and pH treatments resulted in the phenomenon that accommodates this hypothesis; however, resistance development was also observed in this study for strains displaying wild-type status. In alternative to resistance acquisition through horizontal gene transfer, a process known to be modulated by temperature conditions [42], de novo mutations (including recombination) can explain antibiotic resistance development in the absence of resistance determinants or antimicrobial pressure in the environment [45,46], as in this study. In fact, increasing temperatures have been associated with genome-wide selection of these mutations [47]. Despite the costs in fitness already described for resistance acquisition, bacterial species have the potential to downplay such costs by means of compensatory evolution by developing mutations that will decrease fitness cost without compromising antimicrobial resistance or by performing physiological adaptations or activating specific systems that buffer mutational effects and fitness costs [45,46,48,49]. Different factors can influence the acquisition of antibiotic resistance in these settings, such as thermal stress or changes in pH [50,51]. Antimicrobial resistance development occurred in this study for several combinations of water temperature and pH treatments. It is likely that the final antimicrobial susceptibility of the isolates corresponds to an "arms race" between external stressors impact, fitness costs and genetic adaptation by the bacteria, unraveling a non-linear relationship between the tested variables and the antimicrobial susceptibility of *Aeromonas* spp.

4. Materials and Methods

4.1. Strain Selection

Aeromonas species selection followed results obtained prior to this study [52]. Namely, the occurrence of mesophilic *Aeromonas* spp. was investigated in *Iberochondrostoma lusitanicum* in four freshwater streams in the Lisbon district, Portugal (Lizandro: 38.886701°, −9.298140°; Samarra: 38.894761°, −9.433734°; Jamor: 38.720832°, −9.249696°; Laje: 38.709159°, −9.314079°) previously characterized by our team [53]. *A. caviae*, *A. hydrophila*, *A. media* and *A. veronii* were considered the most abundant species and, hence, included in this study. Strains were selected from a bacterial library evaluated by a RAPD (random amplified polymorphic DNA) technique in order to perform molecular typing and genomic differentiation. Three isolates of each *Aeromonas* species that were not considered clones, originating from different locations, were selected as representatives for inclusion in the study (*n* = 12).

The strains' ability to produce slime was evaluated using a phenotypical assay, Congo Red Agar (22 °C, 72 h), as described before [54]. Only slime-producer strains were selected for inclusion in the study.

Strains were stored in pure cultures in cryovials stored at −80 °C. Prior to their use, resuscitation was performed by transferring 100 µL of each bacterial suspension to 8 mL of Brain Heart Infusion broth (BHIB; VWR, Radnor, PA, USA), incubating for 24 h at 21 °C. After, bacterial suspensions were transferred to solid mediums—BHI agar and Columbia Blood (COS) agar (Biomérieux, Marcy-l'Étoile, France)—and incubated at 21 °C for 24 h. The purity of the cultures was confirmed by macro and microscopic morphology, as well as by Gram staining and phenotypic traits (oxidase production).

4.2. Biofilm Formation Quantification

In order to standardize the number of colony-forming units (CFU) in the suspensions to be used in the quantification of biofilm formation, reference *Aeromonas* strains were selected, namely *A. caviae* ATCC 1976, *A. hydrophila* ATCC 7966, *A. media* ATCC 33907 and *A. veronii* ATCC 35624.

Briefly, reference strains were incubated in BHI agar and COS agar at 21 °C for 24 h. After incubation, for each reference strain, colonies were selected and inoculated in 5 mL of 0.9 % saline solution until adjusting to a turbidity of 0.5 McFarland using a digital densitometer DENSIMAT (Biomérieux, Marcy-l'Étoile, France). After homogenization, serial ten-fold dilutions were performed in 9 mL of 0.9% saline solution (up to 10^{-6}). From each dilution (10^{-4} to 10^{-6}), 100 μL were collected and plated in BHI agar in duplicate, using sterilized glass beads. Plates were incubated at 21 °C up to 48 h. Colonies were counted in both plates and averaged. The number of CFU/mL was calculated using the formula (number of colonies × dilution factor)/volume.

Biofilm formation was performed using the microtiter plate assay and quantification was performed using the crystal violet method, as described before [55,56] with modifications. Bacterial colonies were collected from BHI agar and suspended in 5 mL of 0.9% saline solution until adjusting to a turbidity of 0.5 McFarland. Based on the pre-established average CFU/mL for each *Aeromonas* species, concentrations were adjusted for each strain in order to prepare a final concentration in the wells of the Nunc™ MicroWell™ 96-well plates (ThermoFisher Scientific®, Waltham, MA, USA) of 5×10^5 UFC/mL in a final volume of 200 μL. As culture medium, Tryptic Soy Broth (TSB, VWR, Radnor, PA, USA) supplemented with 0.25% glucose (Millipore®, Merck, Darmstadt, Germany) was used. *A. hydrophila* ATCC 7966 is considered a strong biofilm producer; hence it was selected as a positive control. As a negative control, TSB supplemented with 0.25% glucose was used in six wells in each assay. The microtiter plate was incubated at 21 °C for 48 h.

After incubation, the content of all wells was carefully aspirated to eliminate planktonic forms and the wells were washed three times at room temperature with phosphate-buffered saline (PBS; VWR, Radnor, PA, USA) at pH 7.0. The PBS was discarded after the final wash and the microtiter plate was incubated in an inverted position at 60 °C for 1h, for the adherent cells to fixate. After, 150 μL of 0.25% Hucker crystal violet (diluted in de-ionized water; Merck, Darmstadt, Germany) were added to the wells, followed by incubation at room temperature for 5 min. The stain excess was aspirated, and the microtiter plate rinsed until the rinse was free of stain. The microtiter plate was airdried at room temperature and, once dry, 150 μL of 95% ethanol (NORMAPUR®, VWR, Radnor, PA, USA) were added to each well for solubilization of the stain. The microtiter plate was covered with the lid to avoid ethanol's evaporation and incubated at room temperature for 30 min. After incubation, the optical density (OD) of the microtiter plate was evaluated at 570 nm in a horizontal bidirectional reading using the FLUOstar OPTIMA microplate reader (BMG LABTECH, Ortenberg, Germany). This assay was performed prior and after the microcosm assay to enable further comparisons. In both situations, three replicates were performed for each strain on independent days.

4.3. Antimicrobial Susceptibility Testing

Antimicrobial susceptibility testing was performed using the disk diffusion technique [57]. Guidelines of the Clinical and Laboratory Standards Institute for *Aeromonas salmonicida* testing were followed as reference [58], selected since the testing temperature—22 °C—closely resembles the temperature used for the basal treatment. The following antibiotics (Mastdiscs®, Mast Group, Liverpool, UK) were tested: erythromycin (E, 15 μg), tetracycline (T, 30 μg) and sulfamethoxazole/trimethoprim (TS, 23.75–1.25 μg). Antimicrobial compound choice followed options where epidemiological cut-off values were available. A "wild-type" (WT) phenotype implies isolate susceptibility to the antimicrobial, while a "non-wild-type" (NWT) phenotype implies that the isolate presents resistance mechanisms. *Escherichia coli* ATCC 25922 was used as a quality control. This technique was performed prior and after the microcosm assay to enable further comparisons. One strain from each species was randomly selected to be tested. Only strains from pure culture microcosms (i.e., no strains from mixed cultures microcosms were used) were used to perform the antimicrobial susceptibility testing. The same strain was used prior and after microcosm comparisons. In both situations, 10% of replicates were performed on independent days.

4.4. Microcosm Assay

To evaluate the influence that water temperature and pH might have in the antimicrobial resistance and virulence profiles of *Aeromonas* spp., a microcosm simulation assay was developed. Testing variables (i.e., temperature and pH) were selected based on the expected impact that climatic alterations will have in these two parameters in freshwater ecosystems [5] and on the known influence of these variables on bacterial biofilm formation and resistance acquisition/expression [42,59–62].

Regarding water temperature, four experimental conditions were used. First, a condition representing the current water temperature values was created based on trends in water temperature observed during higher temperature months (July to October) in the Lisbon's District rivers (Cascais, Oeiras and Sintra municipalities) in the period between 1985–2016 and averaged (21 °C) [63]. Only sampling points located far from the river mouth were selected to prevent temperature oscillations related to other water bodies. Similarly, only sampling points with substantial datasets over a wide temporal frame were selected ($n = 6$). Location was selected to match the origin of the bacterial isolates. Additionally, two different 21st-century projections of climate alterations for the period of 2081–2100 establishing different levels of greenhouse gas emissions and atmospheric conditions, air pollutant emissions and land use were selected—representative concentration pathways (RCP) 4.5, representing a scenario of medium stabilization (23.2 °C) and 8.5, representing a scenario of high warming (24.5 °C) [5]. To mimic a scenario of rapid temperature fluctuations, the protocol established by Saarinen, Lindström and Ketola [22] was implemented with modifications to accommodate *Aeromonas* spp. growth conditions and the temperature ranges defined for this study. So, repetitions of 24 h cycles of either 24.5 °C or 21 °C were applied. Additionally, to establish an initial time point to enable comparisons in both the microtiter plate assay and the disk diffusion technique prior and after the microcosm assays, a treatment (T0) mimicking the current water temperature and pH (21 °C, pH 7.61) was included. Contrarily to the other treatments, the strains in T0 were incubated in river water for only 24 h.

Simulations from van Vliet et al. [7] on the correlation between air and river water temperature were used to determine final water temperature conditions for the RCP scenarios. Additionally, river discharge level, which also affects water temperature, was based on simulations by van Vliet et al. [7,14] for the Iberian Peninsula and fixed at decrease levels of 40%.

Regarding water pH, and since this parameter trends in rivers will vary according to demographic and geologic characteristics of the areas adjacent to the river [64,65], both a scenario of acidification and a scenario of alkalization were included. Three conditions were created, two mimicking both previously described scenarios and one establishing the current water pH conditions. Water pH values were established based on trends accessed in the same datasets used for temperature [63]. The treatment established as the current condition was based on the average of the values recorded in the analyzed period (pH 7.61). The acidification scenario was based on the average of the lowest pH values observed in all analyzed rivers (pH 6.31), while the alkalization scenario was based on the average of the highest pH values recorded (pH 8.61). A summary of the experimental conditions used in this study is found in Table 1.

Microcosm experimental setup was adapted from Zhang and Buckling [66] and Cairns et al. [67]. Water preparation was performed as described in Sautour et al. [33]. BHIB was used as an addictive of river's water to act as a nutrient source. This medium was used at a 2.5% concentration to resemble the resource levels found in natural ecosystems.

Briefly, river water collected in a freshwater stream in the Lisbon district (Jamor: 38.720832°, −9.249696°) was filtered using a 0.22 μm Millipore filter (Frilabo, Maia, Portugal) and autoclaved at 121 °C for 20 min. For each water pH condition, BHIB was added to the water and pH adjusted to match the conditions established using a HI-4521 Research Grade pH/ORP/EC Bench Meter (Hanna Instruments, Póvoa de Varzim, Portugal). Bacterial suspensions were prepared by collecting colonies from BHI agar that were suspended in 5 mL of 0.9% saline solution until achieving a turbidity of 0.5 McFarland. Suspensions were prepared in pure cultures and in mixed cultures (with only one strain of each species— *A. caviae*, *A. hydrophila*, *A. media* and *A. veronii*—represented once). Nunc™ MicroWell™ 96-well plates were used to establish the microcosm. In pure culture wells, 200 µL of the respective medium were added, following the addition of 10 µL of the bacterial suspension. In the mixed culture wells, 2.5 µL of each bacterial strain was used. In both situations, bacterial suspensions were prepared in 0.9% saline solution previously according to the established average CFU/mL of the reference strains to achieve a final concentration of 5×10^5 UFC/mL in each well. In the negative control wells, 210 µL of the respective medium was added. Plates were incubated for 6 days in the respective temperature treatment inside an SSI10 SSI10-2 orbital shaking incubator (Shel Lab, Cornelius, NC, USA) at 150 rpm to mimic water turbulence in the natural habitat. Every 48 h of incubation, renewal of the medium was performed by adding 20 µL of the previous culture into a new plate with 180 µL of the respective medium. At the end of each microcosm assay, the OD was read at 570 nm as described before to determine bacterial growth. After reading, 10 µL from each well was transferred into BHI agar, incubated at the respective assay's temperature for 24 h and used for biofilm quantification, antimicrobial susceptibility testing and species confirmation (in the case of the mixed culture wells). The pH values for each assay were validated by randomly selecting bacterial cultures across the three different pH used, as well as the negative controls mediums, and analyzed using Neutralit® pH-indicator paper (Merck, Darmstadt, Germany). Tests were performed immediately after incubation.

Table 1. Experimental conditions used in the microcosm assays. RCP—representative concentration pathway.

Experimental Conditions			
Temperature (°C)		**pH**	
Current	21	Current	7.61
RCP 4.5	23.2	Acidification	6.31
RCP 8.5	24.5	Alkalization	8.61
Fluctuations	21–24.5		

4.5. Aeromonas Species Confirmation in Mixed Culture Wells

Following the microcosm assays, species confirmation in the mixed culture wells was performed. Bacterial colonies with distinct macroscopic morphology in BHI agar were selected and streaked into pure cultures. The purity of the cultures was evaluated by macro and microscopic analysis, and Gram staining and oxidase production were evaluated.

Bacterial genomic DNA was obtained by the boiling method [68]. To achieve species identification, a multiplex PCR protocol previously described [69] was used with some modifications. This protocol targets the identification of the four species included in this study. *A. caviae* ATCC 1976, *A. hydrophila* ATCC 7966, *A. media* ATCC 33907 and *A. veronii* ATCC 35624 were used as positive controls.

Briefly, PCR mixtures were performed in a final volume of 25 µL, composed of: 12.15 µL of Supreme NZYTaq 2 × Green Master Mix (NZYTech, Lisbon, Portugal), 10 µL of PCR-grade water (Sigma-Aldrich, Saint Louis, MO, USA), 0.025 µL (0.05 µM) of primers A-16s, 0.25 µL (0.5 µM) of primers A-cav, 0.1 µL (0.2 µM) of primers A-med, 0.225 µL (0.45 µM) of primers A-hyd, 0.075 µL (0.15 µM) of primers A-Ver; and 1.5 µL of template DNA. Thermocycler conditions included a hot start at 95 °C for 2 min; followed by 6 cycles of denaturation at 94 °C for 40 s, annealing at 68 °C for 50 s and extension at 72 °C for 40 s; and 30 cycles at 94 °C for 40 s, 66 °C for 50 s and 72 °C for 40 s.

Amplification products were resolved by gel electrophoresis using 1.5% (*w/v*) agarose in 1 × TBE Buffer (NZYTech, Lisbon, Portugal). Gels were resolved for 45 min at 90 V and NZYDNA Ladder VI (NZYTech, Lisbon, Portugal) was used as a molecular weight marker. Gels were visualized using a UV light transilluminator. The images were recorded through the Bio-Rad ChemiDoc XRS imaging system (Bio-Rad Laboratories, Hercules, CA, USA).

4.6. Statistical Analysis

Prior to statistical analysis, the influence of the microcosm assay (i.e., other factors than the water temperature and pH conditions) on the biofilm production and antimicrobial resistance profiles was accessed by comparing the results obtained with the treatment T0 and current pH 7.61 (similar water temperature and pH conditions). A coefficient of variation of 25% was set as a breakpoint and calculated individually for each *Aeromonas* species. Minimal and maximal limits were calculated regarding T0 values. Current pH 7.61 values that fell outside the limit were considered significantly different. Replicates of isolates where this situation occurred were excluded from the subsequent analysis due to possible bias (i.e., *A. veronii* #1 3rd replicate, mixed culture #1 2nd and 3rd replicates, mixed culture #2 2nd replicate, mixed culture #3 1st and 2nd replicates). For antimicrobial resistance profiles, a qualitative comparison of the epidemiological cut-off values between the two treatments was performed and no deviations occurred.

Several isolate level response variables were analyzed regarding temperature and pH treatments. Using a factorial ANOVA where it was determined the difference in values regarding T0 treatment and Tukey's multiple comparison test to evaluate differences between treatments, the (1) biofilm production and the (2) bacterial growth were considered. Using a stepwise linear regression and a point-biserial correlation, the influence of the different *Aeromonas* species in mixed cultures on the production of biofilm was considered. Pearson's correlation was calculated between biofilm production and bacterial growth. The statistical analysis was performed using IBM SPSS Statistics version 27 software (IBM Analytics, New York, NY, USA). Graphs were produced using GraphPad Prism® (GraphPad Software, San Diego, CA, USA, version 5.01).

5. Conclusions

Current results show how *Aeromonas* spp. will respond to projected environmental shifts in water temperature and pH. Namely, that temperature increments will have a biphasic effect on *Aeromonas* spp. growth, while this bacterial genus will multiply better in acidic environments. Further, *Aeromonas* spp. biofilm production will be decreased due to temperature oscillations and microbial interactions in mixed cultures. Finally, antimicrobial resistance signatures of *Aeromonas* spp. will vary individually to changing temperature and pH parameters. Although general patterns were observed, it is evident that modulation of the intrinsic bacterial characteristics varies across isolates and that the final expression pattern will be influenced by environmental drivers and individual variability; however, the general patterns determined with this study deepen our knowledge on bacterial alterations expected in aquatic environments, strengthening our awareness and response to future bacterial outbreaks and how to deal with them.

Simplification of experimental settings, such as the approach applied in this study, has the limitation of disregarding the role of many other biotic and abiotic factors that can play a role in bacterial growth and virulence and resistance expression. Additionally, focusing on one bacterial genus to study such interactions is a major limitation of this study, since it fails to represent both the outcomes of a bacterial community that closely resembles natural communities, as well as beneficial and detrimental effects of distinct bacterial strains/species on a particular bacterial strain in focus. Further development of microcosm experiments to accommodate more complex networks of drivers and bacterial communities is required.

Author Contributions: Conceptualization, M.L.G., C.S.-S., J.I.R. and M.O.; methodology, M.L.G., A.P., C.S.-S., J.I.R. and M.O.; software, M.L.G. and A.P.; validation, M.L.G. and A.P.; formal analysis, M.L.G. and A.P.; investigation, M.L.G.; resources, M.L.G. and M.O.; data curation, M.L.G. and A.P.; writing—original draft preparation, M.L.G.; writing—review and editing, M.L.G., A.P., C.S.-S., J.I.R. and M.O.; visualization, M.L.G.; supervision, J.I.R. and M.O.; project administration, M.O.; funding acquisition, M.L.G., J.I.R. and M.O. All authors have read and agreed to the published version of the manuscript.

Funding: This research was supported by CIISA—Centro de Investigação Interdisciplinar em Sanidade Animal, Faculdade de Medicina Veterinária, Universidade de Lisboa, Project UIDB/00276/2020 (funded by FCT—Fundação para a Ciência e Tecnologia IP) and by MARE (MARE-ISPA), MARE/UIDB/MAR/04292/2020 and strategic project MARE/UIDP/MAR/04292/2020 (also funded by FCT). MLG thanks funding by the University of Lisbon (PhD fellowship C10571K).

Acknowledgments: The authors wish to thank Eva Cunha and Maria de Fátima Santos (FMV-ULisboa) for their assistance during laboratory work.

Conflicts of Interest: The authors declare no conflict of interest. The funders had no role in the design of the study; in the collection, analyses, or interpretation of data; in the writing of the manuscript, or in the decision to publish the results.

References

1. Parter, M.; Kashtan, N.; Alon, U. Environmental variability and modularity of bacterial metabolic networks. *BMC Evol. Biol.* **2007**, *7*, 169. [CrossRef]
2. Ratzke, C.; Gore, J. Modifying and reacting to the environmental pH can drive bacterial interactions. *PLoS Biol.* **2018**, *16*, e2004248. [CrossRef]
3. De Silva, P.M.; Chong, P.; Fernando, D.M.; Westmacott, G.; Kumar, A. Effect of incubation temperature on antibiotic resistance and virulence factors of *Acinetobacter baumannii* ATCC 17978. *Antimicrob. Agents Chemother.* **2018**, *62*, e01514-17. [CrossRef]
4. Huang, L.; Liu, W.; Jiang, Q.; Zuo, Y.; Su, Y.; Zhao, L.; Qin, Y.; Yan, Q. Integration of transcriptomic and proteomic approaches reveals the temperature-dependent virulence of *Pseudomonas plecoglossicida*. *Front. Cell. Infect. Microbiol.* **2018**, *8*, 207. [CrossRef]
5. Pachauri, R.K.; Allen, M.R.; Barros, V.R.; Broome, J.; Cramer, W.; Christ, R.; Church, J.A.; Clarke, L.; Dahe, Q.; Dasgupta, P.; et al. *Climate Change 2014: Synthesis Report. Contribution of Working Groups I, II and III to the Fifth Assessment Report of the Intergovernmental Panel on Climate Change*; Pachauri, R., Meyer, L., Eds.; IPCC: Geneva, Switzerland, 2014; p. 151.
6. Knouft, J.H.; Ficklin, D. The potential impacts of climate change on biodiversity in flowing freshwater systems. *Annu. Rev. Ecol. Evol. Syst.* **2017**, *48*, 111–133. [CrossRef]
7. Van Vliet, M.T.H.; Ludwig, F.; Zwolsman, J.J.G.; Weedon, G.P.; Kabat, P. Global river temperatures and sensitivity to atmospheric warming and changes in river flow. *Water Resour. Res.* **2011**, *47*. [CrossRef]
8. Bucak, T.; Trolle, D.; Andersen, H.E.; Thodsen, H.; Erdoğan, S.; Levi, E.E.; Filiz, N.; Jeppesen, E.; Beklioğlu, M. Future water availability in the largest freshwater Mediterranean lake is at great risk as evidenced from simulations with the SWAT model. *Sci. Total Environ.* **2017**, *581–582*, 413–425. [CrossRef] [PubMed]
9. Griffith, A.W.; Gobler, C.J. Harmful algal blooms: A climate change co-stressor in marine and freshwater ecosystems. *Harmful Algae* **2020**, *91*, 101590. [CrossRef] [PubMed]
10. Kaushal, S.S.; Likens, G.E.; Pace, M.L.; Utz, R.; Haq, S.; Gorman, J.; Grese, M. Freshwater salinization syndrome on a continental scale. *Proc. Natl. Acad. Sci. USA* **2018**, *115*, E574–E583. [CrossRef]
11. Le, T.D.H., Kattwinkel, M., Schützenmeister, K., Olson, J.R., Hawkins, C.P., Schäfer, R.B. Predicting current and future background ion concentrations in German surface water under climate change. *Philos. Trans. R. Soc. B Biol. Sci.* **2019**, *374*, 20180004. [CrossRef] [PubMed]
12. Dudgeon, D. Multiple threats imperil freshwater biodiversity in the Anthropocene. *Curr. Biol.* **2019**, *29*, R960–R967. [CrossRef]
13. Mosley, L. Drought impacts on the water quality of freshwater systems; review and integration. *Earth Sci. Rev.* **2015**, *140*, 203–214. [CrossRef]
14. Van Vliet, M.T.; Franssen, W.H.; Yearsley, J.R.; Ludwig, F.; Haddeland, I.; Lettenmaier, D.P.; Kabat, P. Global river discharge and water temperature under climate change. *Glob. Environ. Chang.* **2013**, *23*, 450–464. [CrossRef]
15. Pinceel, T.; Buschke, F.; Weckx, M.; Brendonck, L.; Vanschoenwinkel, B. Climate change jeopardizes the persistence of freshwater zooplankton by reducing both habitat suitability and demographic resilience. *BMC Ecol.* **2018**, *18*, 2. [CrossRef] [PubMed]
16. Rodell, M.; Famiglietti, J.; Wiese, D.N.; Reager, J.T.; Beaudoing, H.K.; Landerer, F.W.; Lo, M.-H. Emerging trends in global freshwater availability. *Nature* **2018**, *557*, 651–659. [CrossRef] [PubMed]
17. Chen, H.; Jing, L.; Teng, Y.; Wang, J. Characterization of antibiotics in a large-scale river system of China: Occurrence pattern, spatiotemporal distribution and environmental risks. *Sci. Total Environ.* **2018**, *618*, 409–418. [CrossRef] [PubMed]
18. Peterson, E.; Kaur, P. Antibiotic resistance mechanisms in bacteria: Relationships between resistance determinants of antibiotic producers, environmental bacteria, and clinical pathogens. *Front. Microbiol.* **2018**, *9*, 2928. [CrossRef] [PubMed]

19. Alexander, J.; Bollmann, A.; Seitz, W.; Schwartz, T. Microbiological characterization of aquatic microbiomes targeting taxonomical marker genes and antibiotic resistance genes of opportunistic bacteria. *Sci. Total Environ.* **2015**, *512–513*, 316–325. [CrossRef] [PubMed]

20. Dias, M.F.; da Rocha Fernandes, G.; de Paiva, M.C.; Salim, A.C.D.M.; Santos, A.B.; Nascimento, A.M.A. Exploring the resistome, virulome and microbiome of drinking water in environmental and clinical settings. *Water Res.* **2020**, *174*, 115630. [CrossRef]

21. Friman, V.-P.; Hiltunen, T.; Jalasvuori, M.; Lindstedt, C.; Laanto, E.; Örmälä, A.-M.; Laakso, J.; Mappes, J.; Bamford, J.K.H. High Temperature and bacteriophages can indirectly select for bacterial pathogenicity in environmental reservoirs. *PLoS ONE* **2011**, *6*, e17651. [CrossRef] [PubMed]

22. Saarinen, K.; Lindström, L.; Ketola, T. Invasion triple trouble: Environmental fluctuations, fluctuation-adapted invaders and fluctuation-mal-adapted communities all govern invasion success. *BMC Evol. Biol.* **2019**, *19*, 42. [CrossRef] [PubMed]

23. Abbott, S.L.; Cheung, W.K.; Janda, J.M. The genus aeromonas: Biochemical characteristics, atypical reactions, and phenotypic identification schemes. *J. Clin. Microbiol.* **2003**, *41*, 2348–2357. [CrossRef] [PubMed]

24. Wang, Y.; Gu, J. Influence of temperature, salinity and pH on the growth of environmental *Aeromonas* and *Vibrio* species isolated from Mai Po and the Inner Deep Bay Nature Reserve Ramsar Site of Hong Kong. *J. Basic Microbiol.* **2005**, *45*, 83–93. [CrossRef] [PubMed]

25. Awan, F.; Dong, Y.; Wang, N.; Liu, J.; Ma, K.; Liu, Y. The fight for invincibility: Environmental stress response mechanisms and *Aeromonas hydrophila*. *Microb. Pathog.* **2018**, *116*, 135–145. [CrossRef]

26. Thomas, L.V.; Wimpenny, J.W. Competition between *Salmonella* and *Pseudomonas* species growing in and on agar, as affected by pH, sodium chloride concentration and temperature. *Int. J. Food Microbiol.* **1996**, *29*, 361–370. [CrossRef]

27. Lopez-Vazquez, C.M.; Oehmen, A.; Hooijmans, C.M.; Brdjanovic, D.; Gijzen, H.J.; Yuan, Z.; van Loosdrecht, M.C. Modeling the PAO-GAO competition: Effects of carbon source, pH and temperature. *Water Res.* **2009**, *43*, 450–462. [CrossRef] [PubMed]

28. Yuan, S.; Meng, F. Ecological insights into the underlying evolutionary patterns of biofilm formation from biological wastewater treatment systems: Red or black queen hypothesis? *Biotechnol. Bioeng.* **2020**, *117*, 1270–1280. [CrossRef]

29. Delamare, A.P.L.; Costa, S.O.P.; Da Silveira, M.M.; Echeverrigaray, S. Growth of *Aeromonas* species on increasing concentrations of sodium chloride. *Lett. Appl. Microbiol.* **2000**, *30*, 57–60. [CrossRef]

30. Nana, G.Y.G.; Ripoll, C.; Cabin-Flaman, A.; Gibouin, D.; Delaune, A.; Janniere, L.; Grancher, G.; Chagny, G.; Loutelier-Bourhis, C.; Lentzen, E.; et al. Division-based, growth rate diversity in bacteria. *Front. Microbiol.* **2018**, *9*, 849. [CrossRef]

31. Goelzer, A.; Fromion, V. Bacterial growth rate reflects a bottleneck in resource allocation. *Biochim. Biophys. Acta* **2011**, *1810*, 978–988. [CrossRef]

32. Cavari, B.Z.; Allen, D.; Colwell, R.R. Effect of temperature on growth and activity of *Aeromonas* spp. and mixed bacterial populations in the Anacostia river. *Appl. Environ. Microbiol.* **1981**, *41*, 1052–1054. [CrossRef]

33. Sautour, M.; Mary, P.; Chihib, N.; Hornez, J. The effects of temperature, water activity and pH on the growth of *Aeromonas hydrophila* and on its subsequent survival in microcosm water. *J. Appl. Microbiol.* **2003**, *95*, 807–813. [CrossRef] [PubMed]

34. Palumbo, S.A.; Morgan, D.R.; Buchanan, R.L. Influence of temperature, NaCl, and pH on the growth of *Aeromonas hydrophila*. *J. Food Sci.* **1985**, *50*, 1417–1421. [CrossRef]

35. Knøchel, S. Growth characteristics of motile *Aeromonas* spp. isolated from different environments. *Int. J. Food Microbiol.* **1990**, *10*, 235–244. [CrossRef]

36. Vivekanandhan, G.; Savithamani, K.; Lakshmanaperumalsamy, P. Influence of pH, salt concentration and temperature on the growth of *Aeromonas hydrophila*. *J. Environ. Biol.* **2003**, *24*, 373–379. [PubMed]

37. Karem, K.L.; Foster, J.W.; Bej, A.K. Adaptive acid tolerance response (ATR) in *Aeromonas hydrophila*. *Microbiology* **1994**, *140*, 1731–1736. [CrossRef]

38. Buncic, S.; Avery, S.M. Effect of pre-incubation pH on the growth characteristics of *Aeromonas hydrophila* at 5 °C, as assessed by two methods. *Lett. Appl. Microbiol.* **1995**, *20*, 7–10. [CrossRef]

39. Alvarez-Uria, G.; Midde, M. Trends and factors associated with antimicrobial resistance of *Acinetobacter* spp. invasive isolates in Europe: A country-level analysis. *J. Glob. Antimicrob. Resist.* **2018**, *14*, 29–32. [CrossRef] [PubMed]

40. Kaba, H.E.; Kuhlmann, E.; Scheithauer, S. Thinking outside the box: Association of antimicrobial resistance with climate warming in Europe—A 30 country observational study. *Int. J. Hyg. Environ. Health* **2020**, *223*, 151–158. [CrossRef]

41. Reverter, M.; Sarter, S.; Caruso, D.; Avarre, J.-C.; Combe, M.; Pepey, E.; Pouyaud, L.; Vega-Heredía, S.; De Verdal, H.; Gozlan, R.E. Aquaculture at the crossroads of global warming and antimicrobial resistance. *Nat. Commun.* **2020**, *11*, 70. [CrossRef]

42. MacFadden, D.R.; McGough, S.F.; Fisman, D.; Santillana, M.; Brownstein, J.S. Antibiotic resistance increases with local temperature. *Nat. Clim. Chang.* **2018**, *8*, 510–514. [CrossRef] [PubMed]

43. Andersson, D.I.; Hughes, D. Antibiotic resistance and its cost: Is it possible to reverse resistance? *Nat. Rev. Microbiol.* **2010**, *8*, 260–271. [CrossRef]

44. Sundqvist, M. Reversibility of antibiotic resistance. *Upsala J. Med Sci.* **2014**, *119*, 142–148. [CrossRef]

45. Melnyk, A.H.; Wong, A.; Kassen, R. The fitness costs of antibiotic resistance mutations. *Evol. Appl.* **2015**, *8*, 273–283. [CrossRef]

46. Hernando-Amado, S.; Sanz-García, F.; Blanco, P.; Martínez, J.L. Fitness costs associated with the acquisition of antibiotic resistance. *Essays Biochem.* **2017**, *61*, 37–48. [CrossRef] [PubMed]

47. Berger, D.; Stångberg, J.; Baur, J.; Walters, R.J. Elevated temperature increases genome-wide selection on *de novo* mutations. *Proc. R. Soc. B* **2021**, *288*, 20203094. [CrossRef] [PubMed]

48. Freihofer, P.; Akbergenov, R.; Teo, Y.; Juskeviciene, R.; Andersson, D.I.; Böttger, E.C. Nonmutational compensation of the fitness cost of antibiotic resistance in mycobacteria by overexpression of tlyA rRNA methylase. *RNA* **2016**, *22*, 1836–1843. [CrossRef] [PubMed]
49. Fay, A.; Philip, J.; Saha, P.; Hendrickson, R.C.; Glickman, M.S.; Burns-Huang, K. The DnaK chaperone system buffers the fitness cost of antibiotic resistance mutations in mycobacteria. *mBio* **2021**, *12*, e00123-21. [CrossRef] [PubMed]
50. McMahon, M.A.S.; Xu, J.; Moore, J.E.; Blair, I.S.; McDowell, D.A. Environmental stress and antibiotic resistance in food-related pathogens. *Appl. Environ. Microbiol.* **2007**, *73*, 211–217. [CrossRef]
51. Rodríguez-Verdugo, A.; Gaut, B.S.; Tenaillon, O. Evolution of *Escherichia coli* rifampicin resistance in an antibiotic-free environment during thermal stress. *BMC Evol. Biol.* **2013**, *13*, 50. [CrossRef]
52. Grilo, M.; Isidoro, S.; Chambel, L.; Marques, C.; Marques, T.; Sousa-Santos, C.; Robalo, J.; Oliveira, M. Molecular epidemiology, virulence traits and antimicrobial resistance signatures of *Aeromonas* spp. in the critically endangered *Iberochondrostoma lusitanicum* follow geographical and seasonal patterns. *Antibiotics* **2021**, *10*, 759. [CrossRef]
53. Sousa-Santos, C.; Robalo, J.I.; Pereira, A.M.; Branco, P.; Santos, J.M.; Ferreira, M.T.; Sousa, M.; Doadrio, I. Broad-scale sampling of primary freshwater fish populations reveals the role of intrinsic traits, inter-basin connectivity, drainage area and latitude on shaping contemporary patterns of genetic diversity. *PeerJ* **2016**, *4*, e1694. [CrossRef] [PubMed]
54. Freeman, D.J.; Falkiner, F.R.; Keane, C.T. New method for detecting slime production by coagulase negative staphylococci. *J. Clin. Pathol.* **1989**, *42*, 872–874. [CrossRef] [PubMed]
55. Stepanović, S.; Vuković, D.; Dakić, I.; Savić, B.; Švabić-Vlahović, M. A modified microtiter-plate test for quantification of staphylococcal biofilm formation. *J. Microbiol. Methods* **2000**, *40*, 175–179. [CrossRef]
56. Stepanović, S.; Vukovic, D.; Hola, V.; Di Bonaventura, G.; Djukić, S.; Ćirković, I.; Ruzicka, F. Quantification of biofilm in microtiter plates: Overview of testing conditions and practical recommendations for assessment of biofilm production by staphylococci. *APMIS* **2007**, *115*, 891–899. [CrossRef]
57. Bauer, A.W.; Kirby, W.M.; Sherris, J.C.; Turck, M. Antibiotic susceptibility testing by a standardized single disk method. *Am. J. Clin. Pathol.* **1966**, *45*, 493–496. [CrossRef] [PubMed]
58. Clinical and Laboratory Standards Institute—CLSI. *Performance Standards for Antimicrobial Susceptibility Testing of Bacteria Isolated From Aquatic Animals*; 2nd Informational Supplement; CLSI document VET03/VET04-S2; Clinical and Laboratory Standards Institute: Wayne, PA, USA, 2014.
59. Goller, C.C.; Romeo, T. Environmental influences on biofilm development. In *Bacterial biofilms. Current Topics in Microbiology and Immunology*; Romeo, T., Ed.; Springer: Berlin, Germany, 2008; p. 322.
60. Toyofuku, M.; Inaba, T.; Kiyokawa, T.; Obana, N.; Yawata, Y.; Nomura, N. Environmental factors that shape biofilm formation. *Biosci. Biotechnol. Biochem.* **2016**, *80*, 7–12. [CrossRef]
61. Cruz-Loya, M.; Kang, T.M.; Lozano, N.A.; Watanabe, R.; Tekin, E.; Damoiseaux, R.; Savage, V.M.; Yeh, P.J. Stressor interaction networks suggest antibiotic resistance co-opted from stress responses to temperature. *ISME J.* **2018**, *13*, 12–23. [CrossRef]
62. Mueller, E.; Egan, A.J.; Breukink, E.; Vollmer, W.; Levin, P.A. Plasticity of *Escherichia coli* cell wall metabolism promotes fitness and antibiotic resistance across environmental conditions. *eLife* **2019**, *8*, e40754. [CrossRef]
63. SNIRH. Sistema Nacional de Informação de Recursos Hídricos. 2021. Available online: https://snirh.apambiente.pt/index.php?idMain=2&idItem=1 (accessed on 22 June 2020). (In Portuguese)
64. Kaushal, S.S.; Likens, G.E.; Utz, R.M.; Pace, M.L.; Grese, M.; Yepsen, M. Increased river alkalinization in the Eastern U.S. *Environ. Sci. Technol.* **2013**, *47*, 10302–10311. [CrossRef]
65. Oberholster, P.; Botha, A.-M.; Hill, L.; Strydom, W. River catchment responses to anthropogenic acidification in relationship with sewage effluent: An ecotoxicology screening application. *Chemosphere* **2017**, *189*, 407–417. [CrossRef]
66. Zhang, Q.-G.; Buckling, A. Phages limit the evolution of bacterial antibiotic resistance in experimental microcosms. *Evol. Appl.* **2012**, *5*, 575–582. [CrossRef] [PubMed]
67. Cairns, J.; Becks, L.; Jalasvuori, M.; Hiltunen, T. Sublethal streptomycin concentrations and lytic bacteriophage together promote resistance evolution. *Philos. Trans. R. Soc. B* **2017**, *372*, 20160040. [CrossRef] [PubMed]
68. Talon, D.; Mulin, B.; Thouverez, M. Clonal identification of *Aeromonas hydrophila* strains using randomly amplified polymorphic DNA analysis. *Eur. J. Epidemiol.* **1998**, *14*, 305–310. [CrossRef]
69. Persson, S.; Al-Shuweli, S.; Yapici, S.; Jensen, J.N.; Olsen, K.E.P. Identification of clinical *Aeromonas* species by *rpoB* and *gyrb* sequencing and development of a multiplex PCR method for detection of *Aeromonas hydrophila*, *A. caviae*, *A. veronii*, and *A. media*. *J. Clin. Microbiol.* **2014**, *53*, 653–656. [CrossRef] [PubMed]

Article

Selection of a Gentamicin-Resistant Variant Following Polyhexamethylene Biguanide (PHMB) Exposure in *Escherichia coli* Biofilms

Clémence Cuzin [1], Paméla Houée [1], Pierrick Lucas [2], Yannick Blanchard [2], Christophe Soumet [1] and Arnaud Bridier [1,*]

[1] Antibiotics, Biocides, Residues and Resistance Unit, Fougères Laboratory, French Agency for Food, Environmental and Occupational Health & Safety (ANSES), 35300 Fougères, France; clemence.cuzin@gmail.com (C.C.); pamela.houee@anses.fr (P.H.); christophe.soumet@anses.fr (C.S.)

[2] Viral Genetics and Biosecurity Unit, Ploufragan-Plouzané-Niort Laboratory, French Agency for Food, Environmental and Occupational Health & Safety (ANSES), 22440 Ploufragan, France; pierrick.lucas@anses.fr (P.L.); Yannick.blanchard@anses.fr (Y.B.)

* Correspondence: arnaud.bridier@anses.fr

Citation: Cuzin, C.; Houée, P.; Lucas, P.; Blanchard, Y.; Soumet, C.; Bridier, A. Selection of a Gentamicin-Resistant Variant Following Polyhexamethylene Biguanide (PHMB) Exposure in *Escherichia coli* Biofilms. *Antibiotics* **2021**, *10*, 553. https://doi.org/10.3390/antibiotics10050553

Academic Editors: Manuela Oliveira and Elisabete Silva

Received: 14 April 2021
Accepted: 7 May 2021
Published: 10 May 2021

Abstract: Antibiotic resistance is one of the most important issues facing modern medicine. Some biocides have demonstrated the potential of selecting resistance to antibiotics in bacteria, but data are still very scarce and it is important to better identify the molecules concerned and the underlying mechanisms. This study aimed to assess the potential of polyhexamethylene biguanide (PHMB), a widely used biocide in a variety of sectors, to select antibiotic resistance in *Escherichia coli* grown in biofilms. Biofilms were grown on inox coupons and then exposed daily to sublethal concentrations of PHMB over 10 days. Antibiotic-resistant variants were then isolated and characterized phenotypically and genotypically to identify the mechanisms of resistance. Repeated exposure to PHMB led to the selection of an *E. coli* variant (Ec04m1) with stable resistance to gentamycin (8-fold increase in minimum inhibitory concentration (MIC) compared to the parental strain. This was also associated with a significant decrease in the growth rate in the variant. Sequencing and comparison of the parental strain and Ec04m1 whole genomes revealed a nonsense mutation in the *aceE* gene in the variant. This gene encodes the pyruvate dehydrogenase E1 component of the pyruvate dehydrogenase (PDH) complex, which catalyzes the conversion of pyruvate to acetyl-CoA and CO_2. A growth experiment in the presence of acetate confirmed the role of this mutation in a decreased susceptibility to both PHMB and gentamicin (GEN) in the variant. This work highlights the potential of PHMB to select resistance to antibiotics in bacteria, and that enzymes of central metabolic pathways should be considered as a potential target in adaptation strategies, leading to cross-resistance toward biocides and antibiotics in bacteria.

Keywords: biocide; antibiotic resistance; cross-resistance; aminoglycoside; adaptation; biofilm; pyruvate cycle

1. Introduction

In recent years, the dramatic rise of antimicrobial resistance (AMR) in bacteria has emerged as one of the most challenging concerns for global health. Identifying the drivers of AMR, especially anthropogenic ones, is thus of prime importance to better control the emergence and spread of resistant bacteria [1,2]. In industries, medical areas, public spaces or at home, biocides are broadly used to avoid the dissemination of pathogenic bacteria and guarantee the microbiological quality of equipment, production surfaces and products. However, an increasing number of studies reported the potential of different biocides to impact antibiotic susceptibility in various bacterial species [3–5]. Indeed, some mechanisms responsible for a lower susceptibility in adapted bacteria following a biocide exposure could be also involved in antibiotic resistance, as for instance an upregulated non-specific efflux mediated by multidrug pumps, one of the most recurrently described [6–8]. The adaptation of bacterial populations to biocides particularly occurs in the presence of sublethal concentrations of biocides which select the most tolerant or resistant subpopulations. Different phenomena are at the origin of these sublethal concentrations in industries or at a hospital, for instance as a result of misuse during biocide application or the presence of interfering organic substances [9,10]. In addition, the development of biofilms on surfaces has an effect on biocide concentrations actually experienced by bacteria. Indeed, biofilms are surface-associated bacterial communities embedded in an extracellular matrix, and often exhibit specific functions compared to planktonic cells such as better resistance to disinfectants [11]. The three-dimensional structure of biofilm is known to be able to hinder biocide penetration to the deeper layers, resulting in biocide concentration gradients across the biofilm depth [12,13]. Internal layers can thus be in the presence of lower biocide concentrations that enable bacteria to survive, and trigger an adaptive response to such sub-inhibitory biocide concentrations.

Polyhexamethylene biguanide (PHMB) has been a commonly used biocide for some considerable time, employed in a wide variety of sectors such as disinfectant and antiseptic formulations for wound therapy, in cosmetics, in poultry production to prevent *Salmonella*, or in swimming pool cleanser [14,15]. Despite this broad utilization, data about its potential to select resistance toward other antimicrobials as antibiotics are very scarce and require novel dedicated studies. In this context, we investigated the effects of repeated exposure to PHMB on the antibiotic susceptibility profiles of bacteria in *E. coli* biofilm.

2. Results

2.1. Colony Morphotype Identification Following PHMB Exposure of Ec04 biofilms

Biofilms of the Ec04 strain on stainless steel coupons were exposed daily over 9 days to 1.5625 mg L^{-1} of PHMB, corresponding to sublethal concentrations. Cells were then recovered and plated on trypticase soy agar (TSA) after dilutions. Two different morphotypes were detected, a small colony variant morphotype (Ec04m1) and the parental colony morphotype (Figure 1A). Both morphotypes displayed identical Enterobacterial Repetitive Intergenic Consensus Polymerase Chain Reaction (ERIC-PCR) profiles, suggesting Ec04m1 is not a contamination (Figure 1B).

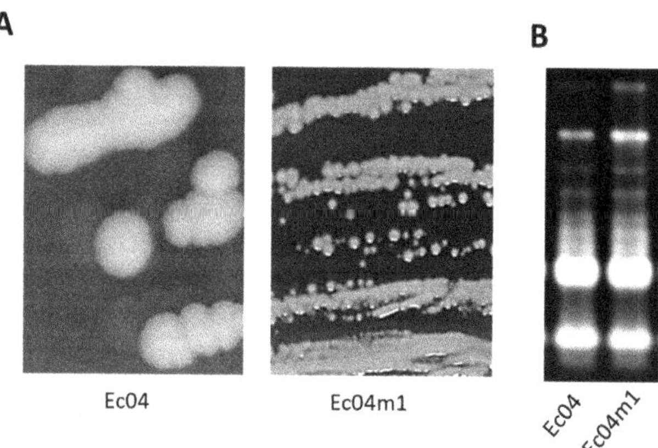

Figure 1. (**A**) Colony morphotypes obtained after plating of Ec04 biofilm exposed to PHMB over 9 days. Ec04m1 corresponds to a small colony variant. (**B**) ERIC-PCR profiles of Ec04 and Ec04m1.

2.2. PHMB and Antibiotic Minimal Inhibitory Concentrations (MIC) Determination for Ec04 Derivatives

Antibiotic and PHMB MIC were determined for the morphotype Ec04m1 in comparison with the parental Ec04. An increase in MIC was observed for Ec04m1 for PHMB (2-fold increase) and two antibiotics, gentamicin (GEN, 8-fold increase) and trimethoprim (TMP, 4-fold increase) (Table 1). These increases led to MIC values of 8 mg L^{-1} and 4 mg L^{-1} for GEN and TMP, respectively, above the ECOFF defined by EUCAST for both antibiotics. A 2-fold decrease in AMP MIC was conversely observed in Ec04m1.

Table 1. Minimal inhibitory concentrations (MIC, mg L^{-1}) for 14 antibiotics and PHMB in Ec04 and derivatives.

Strain	AMP	AZI	FOT	TAZ	CHL	CIP	COL	GEN	MER	NAL	SMX	TET	TIG	TMP	PHMB
Ec04	8	8	0.25	0.5	8	0.015	1	1	0.03	4	16	64	0.25	1	1.5625
Ec04m1	4	8	0.25	0.5	8	0.015	1	8	0.03	4	16	64	0.5	4	3.125
Ec04m1_D2	4	8	0.25	0.5	8	0.03	1	8	0.03	4	32	64	0.5	4	3.125
Ec04m1_D5	2	8	0.25	0.5	8	0.03	1	8	0.03	4	32	64	0.25	4	3.125
Ec04m1_D7	4	8	0.25	0.5	8	0.03	1	8	0.03	4	32	64	0.25	4	3.125
Ec04m1_D10	4	8	0.25	0.5	8	0.03	1	8	0.03	4	32	64	0.25	4	3.125

Ec04m1_D2, Ec04m1_D5, Ec04m1_D7 and Ec04m1_D10 respectively correspond to Ec04m1 after 2, 5, 7 or 10 days of subculture in TSB without PHMB. AMP: ampicillin, AZI: azithromycin, FOT: cefotaxime, CHL: chloramphenicol, CIP: ciprofloxacin, COL: colistin, GEN: gentamicin, MER: meropenem, NAL: nalidixic acid, SMX: sulfamethoxazole, TAZ: ceftazidime, TET: tetracycline, TIG: tigecycline, TMP: trimethoprim.

To assess the stability of this loss of susceptibility toward PHMB, GEN and TMP in the Ec04m1 variant, 10 daily successive subcultures were carried out in a medium without PHMB, and MICs were determined for subcultures at days 2, 5, 7 and 10. The results confirmed the stability of the PHMB, GEN and TMP MIC increase in Ec04m1, as no evolution of MICs was observed after 10 days (Table 1).

2.3. Growth Rate Parameters in Presence of PHMB and GEN

The growth rates of Ec04m1 in gradual concentrations of PHMB or GEN were compared to those obtained from the Ec04 parental strain and are presented in Figure 2. The time to reach the maximal growth rate was also calculated (lag time).

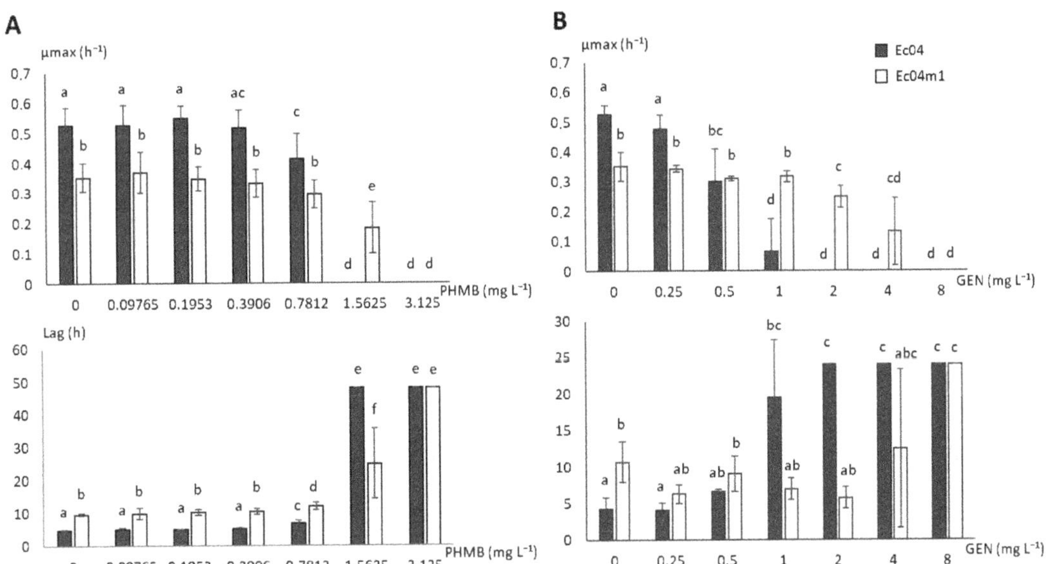

Figure 2. Maximum growth rate (μmax) and lag time (lag) for parental strain Ec04 and Ec04m1 variant in the presence of (**A**) PHMB concentrations from 0 to 3.125 mg L^{-1}, or (**B**) GEN from 0 to 8 mg L^{-1}. Different letters upon the SD bars indicate significant differences between μmax or lag mean values (T-test. $p < 0.05$).

The Ec04m1 variant exhibited a significant decrease in growth rate of 33%, and a 2-fold increase in lag time compared to the parental strain Ec04 in 1/10 TSB medium. A significant decrease ($p < 0.05$) in the µmax value was observed for Ec04 strain from a PHMB concentration of 0.7812 mg L^{-1} (Figure 2A). No significant decrease in the µmax value was observed for Ec04m1 at this PHMB concentration ($p = 0.0968$). Lag times were nevertheless significantly increased for both strains at this concentration. At 1.5625 mg L^{-1} of PHMB, a concentration corresponding to the MIC value determined for Ec04, and no growth was detected for this strain, whereas the Ec04m1 variant was able to grow with a µmax equal to 0.183 h^{-1}. At 3.125 mg L^{-1} of PHMB, growth was inhibited for both strains. These results underlined the ability of the variant Ec04 to survive in higher PHMB concentrations compared to the parental Ec04 strain.

Comparable results were obtained with GEN (Figure 2B). Indeed, the µmax and lag time were deeply affected by GEN from a concentration of 1 mg L^{-1} ($p < 0.05$) in the Ec04 strain, whereas no effect was observed on growth parameters at this concentration for the Ec04m1 variant. While the growth of Ec04 strain is fully inhibited at 2 mg L^{-1}, the µmax obtained for Ec04m1 variant gradually decreased from 2 to 8 mg L^{-1}, confirming its lower susceptibility to this aminoglycoside antibiotic.

2.4. Genomic Characterization of Ec04m1 Variant

The genomes of strains Ec04 (biosample SAMN16976434) and Ec04m1 (SAMN16976489) were sequenced and compared to investigate the genetic origin of the Ec04m1 variant phenotype. Only one SNP with a minimum frequency >80% was identified in the Ec04m1 variant compared to Ec04, leading to the substitution of cytosine by thymine at position 421 in the *aceE* gene (Figure 3A). This mutation results in a stop codon at the location of glutamine, leading to a truncated AceE protein in the Ec04m1 variant (140aa versus 887 in the parental Ec04 strain) (Figure 3B). AceE is the pyruvate dehydrogenase (PDH) E1 component and forms with AceF and LpdA, the PDH enzyme complex normally catalyzing the conversion of pyruvate in acetyl CoA under the regulation of a pyruvate-sensing PdhR regulator. The mutation detected in Ec04m1 thus impairs acetyl-CoA production from pyruvate, and alters a central pathway critical for the tricarboxylic acid (TCA) cycle (Figure 3C).

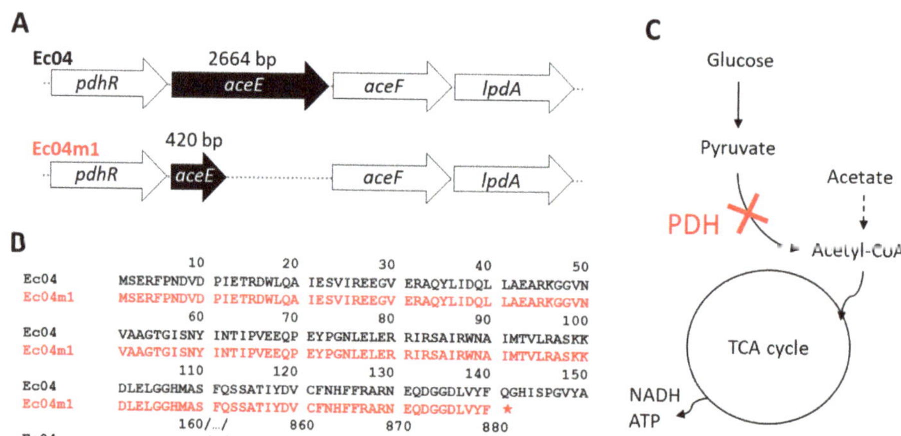

Figure 3. Detection of a nonsense mutation in the Ec04m1 variant in *aceE* gene encoding the pyruvate dehydrogenase E1 component of the pyruvate dehydrogenase (PDH) complex that catalyzes the conversion of pyruvate to acetyl-CoA and CO_2. The mutation (C > T at position 421) results in a premature stop codon at the location of glutamine, leading to a truncated *aceE* protein (140aa). (**A**) Schematic representation of the *aceE* gene and its genetic environment (*pdhR-aceEF-lpdA* operon) in Ec04 and Ec04m1 strains. (**B**) Protein sequence of *aceE* in Ec04 (887aa) and its alignment with the truncated protein in Ec04m1 (140aa); stars symbolize stop codons. (**C**) Schematic representation of the central metabolic pathway altered by the mutation in the Ec04m1 variant.

2.5. Effect of Acetate on Growth Rate in the Presence of GEN and PHMB

The effect of acetate addition in the growth medium on the growth rates of the Ec04m1 variant in the presence of PHMB or GEN was assessed, since alternative pathways for the synthesis of acetyl-CoA from acetate exist in *E.coli*. In the absence of PHMB and GEN, the presence of acetate led to a slight but significant increase in growth rate in the Ec04m1 variant ($p < 0.05$) although it remains below that measured for Ec04 (Figure 4). At a PHMB concentration of 1.5625 mg L^{-1}, Ec04m1 growth was almost totally inhibited in the presence of 30 mM acetate, as in the Ec04 parental strain, while the variant was able to grow in the absence of acetate under identical conditions as previously shown. Similar observations were made in the presence of GEN. As shown in Figure 4, at an antibiotic concentration of 1 and 2 mg L^{-1}, the presence of acetate led to a significant decrease in the Ec04m1 growth rate, although remaining significantly higher than that observed for Ec04. The presence of 30 mM acetate resulted in the inhibition of Ec04m1 growth at a GEN concentration of 4 mg L^{-1}, while Ec04m1 was able to grow at this antibiotic concentration in the absence of acetate. Overall, the supplementation of acetate in the growth medium led to a partial restoration of the susceptibility of the Ec04m1 strain to both PHMB and GEN.

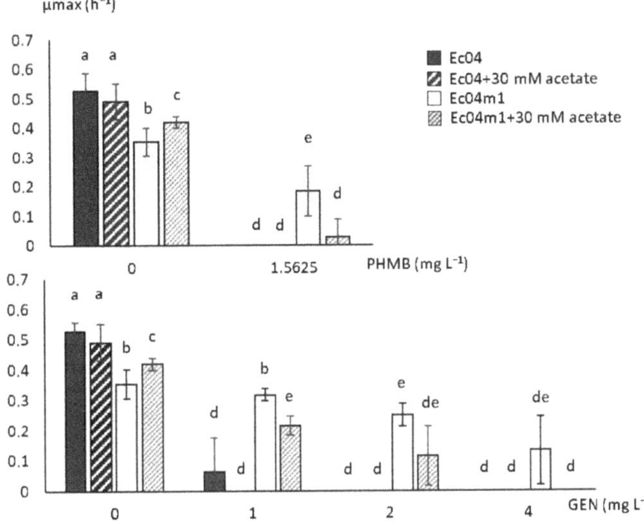

Figure 4. Effect of acetate medium supplementation (30 mM) on maximal growth rate (μmax) for parental strain Ec04 and Ec04m1 variant in the presence of PHMB or GEN. Different letters upon the SD bars indicate significant differences between μmax or lag mean values (T-test. $p < 0.05$).

3. Discussion

Large quantities of biocides are used every day to prevent the spread of pathogenic microorganisms in a wide variety of sectors. An increasing quantity of evidence emphasizes a connection between biocide exposure and the modification of susceptibility toward antibiotics in bacteria [3,4]. A better understanding of the potential of various biocides to select resistance toward antibiotics, along with the associated mechanisms, is therefore required in the perspective to combat antibiotic resistance emergence. In this study, we highlighted the potential of 1.5625 mg L^{-1} PHMB exposure to select resistance to antibiotics, especially GEN, in *E. coli*. This sublethal concentration is approximately 100 times lower than the end-use concentration, for instance when disinfecting hard surfaces (0.16 g L^{-1}, [16]). Previous studies have shown a decrease in susceptibilities toward antibiotics in various bacterial species after exposure to sublethal concentrations of chlorhexidine, another member of the biguanide molecule family [17,18]. Interestingly, Henly et al. [19] reported that PHMB exposure induced resistance to trimethoprim-sulfamethoxazole in the CFT073 *E. coli* strain, and resistance to GEN in the EC26 strain, which is consistent with the results obtained in the present work. Here, the Ec04m1 variant also exhibited a 2-fold increase in PHMB MIC after repeated exposure. This suggests the mechanisms leading to this lower susceptibility due to adaptation to PHMB could be shared with those playing a role in antibiotic resistance. Polymeric biguanides, such as PHMB, first act on bacteria through the interaction with cations associated with the cell envelope, causing membrane destabilization and LPS reorganization [20]. This cellular uptake mechanism is shared with polycationic agents, such as aminoglycosides like GEN, and therefore modifications of membrane permeability in variant Ec04m1, altering cellular uptake, could explain both decreases in susceptibility to PHMB and GEN as proposed previously [19].

WGS revealed the presence of a unique mutation in Ec04m1 compared to the Ec04 parental strain in the *aceE* gene, which encodes the pyruvate dehydrogenase E1 component. The mutation resulted in a truncated protein in the variant, impairing the conversion of pyruvate in Acetyl-CoA, a central pathway in carbon metabolism in *E. coli*. Consistently with the significant decrease in growth rate observed in the Ec04m1 variant in comparison with the Ec04 parental strain (Figure 2), it was previously shown that *aceE* deletion greatly affected growth rate and biosynthetic capacity in *E. coli* [21,22]. Globally, the depletion of enzymes of the pyruvate cycle, such as pyruvate dehydrogenase, has been considered to shut down the TCA cycle, and thus to profoundly affect energy production and regulation in *E. coli*. Interestingly, Schutte et al. [21] showed that the pyruvate dehydrogenase complex is a central component in the antimicrobial activity mediated by the chemokine CXCL10 in *E. coli*. Changes in carbon metabolism activity are also known to affect bacterial susceptibility toward antibiotics including aminoglycosides in *E. coli* [23,24]. Recently, it was also shown that inactivation of central carbon metabolism enzymes can also participate in antibiotic resistance in other Gram-negative species such as *Stenotrophomonas maltophilia* [25]. In line with this, stimulation of the central metabolism using various metabolites enabled researchers to potentiate aminoglycoside efficacy in *E. coli* or *Pseudomonas* spp. through the stimulation of the proton motive force (PMF) favoring the uptake of antibiotics across the bacterial membrane [23,26,27]. As reported by Chindera et al. [14], PHMB also enters the bacterial cell through an energy-dependent uptake process, since authors showed that bacteria cultivated at 4 °C displayed reduced PHMB uptake compared to cells held at 37 °C. The lower susceptibility of the Ec04m1 variant selected upon PHMB exposure could thus be related to the reduced uptake of antimicrobial molecules due to the altered pyruvate cycle and reduced metabolic activity. Moreover, the deletion of *aceE* greatly alters fatty acid biosynthesis, which is linked to the central carbon cycle through the use of acetyl-CoA [22]. This biosynthesis pathway is crucial for lipid synthesis in the cell envelope, emphasizing a relevant connection between the alteration of the central carbon biosynthesis pathway in the Ec04m1 variant and modification of the membrane permeability, which could therefore play a role in decreased susceptibility to both PHMB and GEN, as previously suggested.

To confirm the role of the mutation in *aceE* in the decreased susceptibility to GEN and PHMB, observed in the Ec04m1 variant, growth rate measurements were performed in the presence of 30 mM acetate. Acetate addition has been shown to suppress the effects of *aceE* deletion in *E. coli*, allowing the synthesis of acetyl-CoA by an alternative pathway (Figure 3C) [28]. Our results revealed that acetate growth medium supplementation did indeed result in a significant increase in the growth rate of the Ec04m1 variant and resulted in a partial restoration of its susceptibility to both PHMB and GEN (Figure 4). This observation thus confirmed the involvement of the pyruvate cycle alteration in the adaptation to biocides after repeated exposure, and the development of GEN resistance. While this central metabolism alteration penalizes the Ec04m1 variant in the absence of biocides or antibiotics by decreasing its growth rate, the fact that it was favored at low PHMB concentrations (1.5625 mg L^{-1}) emphasizes its potential ability to survive in environments where residual concentrations of PHMB are present. PHMB, in addition to being used in a wide variety of applications, is a chemically stable molecule, especially in water, and thus can persist for a long time in the environment at these low concentrations [29]. Moreover, the presence of interfering substances or biofilms can reduce biocide concentrations that reach bacteria [12]. Such phenomena are likely to create favorable micro-environmental conditions for the survival and dissemination of the resistant variant, thus representing a risk for public health.

4. Materials and Methods

4.1. Bacterial Strain and Growth Conditions

The *E. coli* strain used in this study (called hereafter Ec04) was initially isolated from a pig caecum in a slaughterhouse (Brittany, France). Bacterial stock cultures were kept at $-80\,°C$ in a cryoprotective solution (0.5% tryptone, 0.3% beef extract, 15% glycerol). Prior to each experiment, frozen cells were firstly sub-cultured on a trypticase soy agar (TSA) plate over 24 h at 37 °C, and a colony was then transferred to tryptone soya broth (TSB) at 37 °C overnight.

Biofilms were grown on sterile $10 \times 20 \times 1\ mm^3$ stainless steel coupons by inoculating wells of a 6-well microtiter plate with 4 mL of overnight bacterial suspension adjusted in 1/10 TSB to 10^3 CFU mL^{-1}. The 6-well microplate was then incubated at 20 °C for 72 h to enable the formation of the biofilm on the coupons at a cell density of approximately 10^7 CFU/cm^2.

4.2. Biocide Susceptibility Testing

PHMB MIC were determined using a microdilution broth method adapted from that of Schug et al. [30]. Briefly, each well of a 6-well microtiter plate was inoculated with 2 mL of biocide at the desired concentration, and 2 mL of a planktonic suspension was obtained after 72 h growth at 20 °C in 1/10 TSB and then adjusted to 10^7 CFU mL^{-1}. Plates were then incubated at 20 °C for 24 h (also checked at 48 h) and MIC was determined as the lowest concentration of biocide that prevents bacterial growth. MIC values were determined through the reading of turbidity and confirmed by drop-plating on TSA plates to confirm the reading through colony growth examination. All the determinations of MIC were repeated at least twice.

4.3. Antibiotic Susceptibility Testing

Antibiotic susceptibility tests were performed using a standard microdilution method (EUVSEC, Sensititre®, TREK Diagnostic Systems Ltd., Thermo Fisher Scientific, East Grinstead, UK), using a panel of 14 antimicrobial substances according to manufacturer instructions. The strains were interpreted as resistant to antibiotics according to the epidemiological resistance cut-off value (ECOFF) determined by EUCAST (European Committee on Antimicrobial Susceptibility Testing, http://mic.eucast.org, accessed on 14 April 2021). *E. coli* ATCC 25,922 was used as quality control. All the determinations of MIC were repeated at least twice, and thrice if values were not similar.

4.4. Adaptation Experiments to PHMB and Identification of Colony Morphology Variants

Biofilms were repeatedly exposed to a sublethal concentration of PHMB at 1.5625 mg L^{-1} corresponding to the MIC obtained for planktonic suspension of the Ec04 strain. Concretely, biofilm coupons were daily transferred into new wells of 6-well microtiter plates filled with fresh 1/10 TSB medium containing PHMB at 1.5625 mg L^{-1}, over 9 days, and then biofilm coupons were transferred in a neutralizing agent (REF + composition). Coupons were then ultra-sonicated and vortexed to recover biofilm cells. Serial dilutions were then performed and cells were plated on TSA. The different morphotypes were then checked on the TSA plate, and each colony with a distinct morphology was isolated and stocked in a cryoprotective medium at $-80\,°C$ before further experiments.

ERIC-PCR was used to compare fingerprint patterns obtained with morphotypes and the Ec04 parental strain to ensure that the different colony morphotypes did indeed correspond to Ec04 derivatives and not to a contamination event. To further this aim, DNA was extracted using an InstaGene kit (Bio-rad, Marnes-la-Coquette, France) and amplified using a LightCycler® 480 thermocycler (Roche Diagnostics, Meylan, France) with primers ERIC1-R (ATGTAAGCTCCTGGGGATTCAC) and ERIC2 (AAGTAAGTGACTGGGGT-GAGCG) [31] and GoTaq Flexi polymerase (Promega, Charbonnières-les-bains, France) as follows: 95 °C for 2 min for initial melting; 30 cycles at 95 °C for 1 min, 54 °C for 1 min, 72 °C for 4 min; final extension at 72 °C for 8 min followed by incubation at 4 °C. PCR products were then checked on 1% agarose gel and migrated over 90 min at 110 V before being revealed using a GelRED stain (Biotium, Brumath, France).

MIC for biocides and antibiotics were determined as described earlier, to evaluate changes in antimicrobial resistance profiles, due to repeated PHMB exposure, for confirmed morphotype variants. In the case of MIC, it increased compared to WT. Daily subcultures over 10 days in fresh TSB without PHMB were performed, and MICs were determined again on the cells after days 2, 5, 7 and 10, to assess the stability of the potential modification of antibiotic susceptibility as observed.

4.5. Growth Parameter Measurements

The growth parameters of bacterial strains were determined by automatically measuring the OD of cultures at 620 nm in a FLUOstar Optima microplate reader (BMG Labtech, Champigny sur Marne, France) in the presence or absence of PHMB or GEN. Briefly, 4 mL of TSB were inoculated from cryotubes and incubated at 37 °C for 24 h. Then, 300 µL from this suspension was transferred in 4 mL TSB and incubated for 6–7 h at 37 °C. The bacterial suspension was adjusted to 10^3 CFU mL^{-1} and used to inoculate the 96 wells of a microtiter plate (Greiner) containing 1/10 TSB (with or without 30 mM acetate) and a range of concentrations of PHMB (0–3.125 mg L^{-1}) or GEN (0–8 mg L^{-1}). OD_{620nm} was then automatically measured every 30 min over 48 h at 37 °C by the FLUOstar reader. Maximum growth rate (μ_{max}) values and lag time (λ) were extracted from OD curves using MARS data analysis software (version 2.10, BMG Labtech, Champigny sur Marne, France). Experiments were performed at least three times independently of duplicates.

4.6. WGS and Mutation Detection

Genomic DNA was extracted from the pellet of a 5 mL exponential culture of strains Ec04 or Ec04m1, grown in TSB centrifuged at 8000 g for 5 min using the Nucleospin tissue kit (Macherey-Nagel, Duren, Germany) according to manufacturer instructions. The quantity and quality of the DNA were checked using a BioSpec-Nano (Shimadzu, Marne la Vallée, France) spectrophotometer. Whole-genome sequencing was performed with the NextSeq 500 (Illumina). Reads were assembled using the Shovill pipeline (v0.9.0, https://github.com/tseemann/shovill (accessed on 30 April 202).). This pipeline used Trimmomatic (v0.38) and SPAdes (v3.13.0). All contigs with a length shorter than 200 nucleotides and a kmer coverage lower than 2 are filtered. Genomes were annotated with Prokka [32] (v1.13.3), then a BWA-MEM [33] (v0.7.8) alignment was performed using the Ec04 genome as a reference against all reads cleaned by Trimmomatic (ILLUMINACLIP: oligos.fasta: 2:30:5:1: true; LEADING: 3; TRAILING: 3; MAXINFO: 40:0.2; MINLEN: 36). Variant calling was performed using the bwa alignment file with VarScan [34] (version 2.4; parameters: min-coverage, 8; min-reads2, 2; min-avg-qual, 15; min-var-freq, 0.2; p value, 0.05; strand-filter disabled; variants 1).

4.7. Sequence Accession Numbers

Whole-genome sequence assemblies have been deposited in NCBI in bio-project PRJNA681999, with bio-sample numbers SAMN16976434 for Ec04 and SAMN16976489 for Ec04m1.

5. Conclusions

This study revealed the potential of PHMB exposure to select resistance toward antibiotics in *E. coli* through the alteration of central carbon flow. These observations emphasize the ability of bacterial populations to adapt to various antimicrobial stresses by reshaping their central metabolism, which might constitute an important target to be considered when understanding cross-resistance emergence between biocides and antibiotics.

Author Contributions: Conceptualization, A.B. and C.S.; methodology, A.B., C.C., P.H., P.L.; validation, A.B. and C.S.; investigation, A.B., C.C., and P.L.; resources, A.B., Y.B., and C.S.; data curation, A.B., C.C., and P.L.; writing—original draft preparation, A.B.; writing—review and editing, A.B., P.L., Y.B., and C.S.; supervision, A.B. and C.S.; project administration, A.B. All authors have read and agreed to the published version of the manuscript.

Funding: This research received no external funding.

Institutional Review Board Statement: Not applicable.

Informed Consent Statement: Not applicable.

Data Availability Statement: All data were available in the manuscript. Sequencing data were deposited in NCBI public database under the bio-project PRJNA681999.

Conflicts of Interest: The authors declare no conflict of interest.

References

1. Vikesland, P.; Garner, E.; Gupta, S.; Kang, S.; Maile-Moskowitz, A.; Zhu, N. Differential Drivers of Antimicrobial Resistance across the World. *Acc. Chem. Res.* **2019**, *52*, 916–924. [CrossRef]
2. Swift, B.M.C.; Bennett, M.; Waller, K.; Dodd, C.; Murray, A.; Gomes, R.L.; Humphreys, B.; Hobman, J.L.; Jones, M.A.; Whitlock, S.E.; et al. Anthropogenic Environmental Drivers of Antimicrobial Resistance in Wildlife. *Sci. Total Environ.* **2019**, *649*, 12–20. [CrossRef]
3. Guérin, A.; Bridier, A.; Le Grandois, P.; Sévellec, Y.; Palma, F.; Félix, B.; Roussel, S.; Soumet, C.; Listadapt Study Group. Exposure to Quaternary Ammonium Compounds Selects Resistance to Ciprofloxacin in *Listeria monocytogenes*. *Pathogens* **2021**, *10*, 220. [CrossRef]
4. Kampf, G. Biocidal Agents Used for Disinfection Can Enhance Antibiotic Resistance in Gram-Negative Species. *Antibiotics* **2018**, *7*, 110. [CrossRef]
5. Westfall, C.; Flores-Mireles, A.L.; Robinson, J.I.; Lynch, A.J.L.; Hultgren, S.; Henderson, J.P.; Levin, P.A. The Widely Used Antimicrobial Triclosan Induces High Levels of Antibiotic Tolerance In Vitro and Reduces Antibiotic Efficacy up to 100-Fold In Vivo. *Antimicrob. Agents Chemother.* **2019**, *63*. [CrossRef]
6. Wales, A.D.; Davies, R.H. Co-Selection of Resistance to Antibiotics, Biocides and Heavy Metals, and Its Relevance to Foodborne Pathogens. *Antibiotics* **2015**, *4*, 567–604. [CrossRef]
7. Amsalu, A.; Sapula, S.A.; De Barros Lopes, M.; Hart, B.J.; Nguyen, A.H.; Drigo, B.; Turnidge, J.; Leong, L.E.; Venter, H. Efflux Pump-Driven Antibiotic and Biocide Cross-Resistance in *Pseudomonas aeruginosa* Isolated from Different Ecological Niches: A Case Study in the Development of Multidrug Resistance in Environmental Hotspots. *Microorganisms* **2020**, *8*, 1647. [CrossRef] [PubMed]
8. Buffet-Bataillon, S.; Tattevin, P.; Maillard, J.-Y.; Bonnaure-Mallet, M.; Jolivet-Gougeon, A. Efflux Pump Induction by Quaternary Ammonium Compounds and Fluoroquinolone Resistance in Bacteria. *Future Microbiol.* **2016**, *11*, 81–92. [CrossRef] [PubMed]
9. Araújo, P.A.; Lemos, M.; Mergulhão, F.; Melo, L.; Simões, M. The Influence of Interfering Substances on the Antimicrobial Activity of Selected Quaternary Ammonium Compounds. *Int. J. Food Sci.* **2013**, *2013*. [CrossRef]
10. Lambert, R.J.; Johnston, M.D. The Effect of Interfering Substances on the Disinfection Process: A Mathematical Model. *J. Appl. Microbiol.* **2001**, *91*, 548–555. [CrossRef] [PubMed]
11. Bridier, A.; Sanchez-Vizuete, P.; Guilbaud, M.; Piard, J.-C.; Naïtali, M.; Briandet, R. Biofilm-Associated Persistence of Food-Borne Pathogens. *Food Microbiol.* **2015**, *45*, 167–178. [CrossRef]
12. Bridier, A.; Briandet, R.; Thomas, V.; Dubois-Brissonnet, F. Resistance of Bacterial Biofilms to Disinfectants: A Review. *Biofouling* **2011**, *27*, 1017–1032. [CrossRef] [PubMed]
13. Bridier, A.; Dubois-Brissonnet, F.; Greub, G.; Thomas, V.; Briandet, R. Dynamics of the Action of Biocides in *Pseudomonas aeruginosa* Biofilms. *Antimicrob. Agents Chemother.* **2011**, *55*, 2648–2654. [CrossRef]
14. Chindera, K.; Mahato, M.; Sharma, A.K.; Horsley, H.; Kloc-Muniak, K.; Kamaruzzaman, N.F.; Kumar, S.; McFarlane, A.; Stach, J.; Bentin, T.; et al. The Antimicrobial Polymer PHMB Enters Cells and Selectively Condenses Bacterial Chromosomes. *Sci. Rep.* **2016**, *6*. [CrossRef]
15. Elsztein, C.; de Lucena, R.M.; de Morais, M.A. The Resistance of the Yeast *Saccharomyces cerevisiae* to the Biocide Polyhexamethylene Biguanide: Involvement of Cell Wall Integrity Pathway and Emerging Role for YAP1. *BMC Mol. Biol.* **2011**, *12*. [CrossRef] [PubMed]
16. ECHA French Assessment Report for the Evaluation of Active Substances in the Frame of Regulation (EU) No 528/2012 Concerning the Making Available on the Market and Use of Biocidal Products. Polyhexamethylene Biguanide PT02. 2017. Available online: https://echa.europa.eu/documents/10162/8e7a0fe2-6b23-1705-d797-85c34e6f205b (accessed on 31 December 2020).

17. Wand, M.E.; Bock, L.J.; Bonney, L.C.; Sutton, J.M. Mechanisms of Increased Resistance to Chlorhexidine and Cross-Resistance to Colistin Following Exposure of *Klebsiella pneumoniae* Clinical Isolates to Chlorhexidine. *Antimicrob. Agents Chemother.* **2017**, *61*. [CrossRef] [PubMed]

18. Kampf, G. Acquired Resistance to Chlorhexidine-Is It Time to Establish an "antiseptic Stewardship" Initiative? *J. Hosp. Infect.* **2016**, *94*, 213–227. [CrossRef] [PubMed]

19. Henly, E.L.; Dowling, J.A.R.; Maingay, J.B.; Lacey, M.M.; Smith, T.J.; Forbes, S. Biocide Exposure Induces Changes in Susceptibility, Pathogenicity, and Biofilm Formation in Uropathogenic *Escherichia coli*. *Antimicrob. Agents Chemother.* **2019**, *63*. [CrossRef]

20. Gilbert, P.; Allison, D.G.; McBain, A.J. Biofilms in Vitro and in Vivo: Do Singular Mechanisms Imply Cross-Resistance? *J. Appl. Microbiol.* **2002**, *92*, 98S–110S. [CrossRef]

21. Schutte, K.M.; Fisher, D.J.; Burdick, M.D.; Mehrad, B.; Mathers, A.J.; Mann, B.J.; Nakamoto, R.K.; Hughes, M.A. *Escherichia coli* Pyruvate Dehydrogenase Complex Is an Important Component of CXCL10-Mediated Antimicrobial Activity. *Infect. Immun.* **2016**, *84*, 320–328. [CrossRef] [PubMed]

22. Westfall, C.S.; Levin, P.A. Comprehensive Analysis of Central Carbon Metabolism Illuminates Connections between Nutrient Availability, Growth Rate, and Cell Morphology in *Escherichia coli*. *PLoS Genet.* **2018**, *14*. [CrossRef]

23. Su, Y.-B.; Peng, B.; Li, H.; Cheng, Z.-X.; Zhang, T.-T.; Zhu, J.-X.; Li, D.; Li, M.-Y.; Ye, J.-Z.; Du, C.-C.; et al. Pyruvate Cycle Increases Aminoglycoside Efficacy and Provides Respiratory Energy in Bacteria. *Proc. Natl. Acad. Sci. USA* **2018**, *115*, E1578–E1587. [CrossRef] [PubMed]

24. Yang, Y.; Mi, J.; Liang, J.; Liao, X.; Ma, B.; Zou, Y.; Wang, Y.; Liang, J.; Wu, Y. Changes in the Carbon Metabolism of *Escherichia coli* During the Evolution of Doxycycline Resistance. *Front. Microbiol.* **2019**, *10*. [CrossRef]

25. Gil-Gil, T.; Corona, F.; Martínez, J.L.; Bernardini, A. The Inactivation of Enzymes Belonging to the Central Carbon Metabolism Is a Novel Mechanism of Developing Antibiotic Resistance. *mSystems* **2020**, *5*. [CrossRef] [PubMed]

26. Crabbé, A.; Ostyn, L.; Staelens, S.; Rigauts, C.; Risseeuw, M.; Dhaenens, M.; Daled, S.; Van Acker, H.; Deforce, D.; Van Calenbergh, S.; et al. Host Metabolites Stimulate the Bacterial Proton Motive Force to Enhance the Activity of Aminoglycoside Antibiotics. *PLoS Pathog.* **2019**, *15*. [CrossRef]

27. Martins, D.; Nguyen, D. Stimulating Central Carbon Metabolism to Re-Sensitize *Pseudomonas aeruginosa* to Aminoglycosides. *Cell. Chem. Biol.* **2017**, *24*, 122–124. [CrossRef] [PubMed]

28. Battesti, A.; Majdalani, N.; Gottesman, S. Stress Sigma Factor RpoS Degradation and Translation Are Sensitive to the State of Central Metabolism. *Proc. Natl. Acad. Sci. USA* **2015**, *112*, 5159–5164. [CrossRef] [PubMed]

29. Lucas, A.D. Environmental Fate of Polyhexamethylene Biguanide. *Bull. Environ. Contam. Toxicol.* **2012**, *88*, 322–325. [CrossRef]

30. Schug, A.R.; Bartel, A.; Scholtzek, A.D.; Meurer, M.; Brombach, J.; Hensel, V.; Fanning, S.; Schwarz, S.; Feßler, A.T. Biocide Susceptibility Testing of Bacteria: Development of a Broth Microdilution Method. *Vet. Microbiol.* **2020**, *248*. [CrossRef] [PubMed]

31. Hulton, C.S.; Higgins, C.F.; Sharp, P.M. ERIC Sequences: A Novel Family of Repetitive Elements in the Genomes of *Escherichia coli*, *Salmonella typhimurium* and Other Enterobacteria. *Mol. Microbiol.* **1991**, *5*, 825–834. [CrossRef]

32. Seemann, T. Prokka: Rapid Prokaryotic Genome Annotation. *Bioinformatics* **2014**, *30*, 2068–2069. [CrossRef] [PubMed]

33. Li, H.; Durbin, R. Fast and Accurate Short Read Alignment with Burrows-Wheeler Transform. *Bioinformatics* **2009**, *25*, 1754–1760. [CrossRef] [PubMed]

34. Koboldt, D.C.; Zhang, Q.; Larson, D.E.; Shen, D.; McLellan, M.D.; Lin, L.; Miller, C.A.; Mardis, E.R.; Ding, L.; Wilson, R.K. VarScan 2: Somatic Mutation and Copy Number Alteration Discovery in Cancer by Exome Sequencing. *Genome Res.* **2012**, *22*, 568–576. [CrossRef] [PubMed]

MDPI

St. Alban-Anlage 66

4052 Basel

Switzerland

Tel. +41 61 683 77 34

Fax +41 61 302 89 18

www.mdpi.com

Antibiotics Editorial Office

E-mail: antibiotics@mdpi.com

www.mdpi.com/journal/antibiotics